HOW TO BECOME A Top SAS® Programmer

Michael A. Raithel

support.sas.com/publishing

The correct bibliographic citation for this manual is as follows: Raithel, Michael A. 2013. *How to Become a Top SAS® Programmer*. Cary, NC: SAS Institute Inc.

How to Become a Top SAS® Programmer

ISBN 978-4-61290-633-1 (electronic book)
ISBN 978-1-61290-104-6

SAS Institute Inc., SAS Campus Drive, Cary, North Carolina 27513-2414.

Printing 1, August 2013

SAS provides a complete selection of books and electronic products to help customers use SAS® software to its fullest potential. For more information about our offerings, visit **support.sas.com/bookstore** or call 1-800-727-3228.

Contents

About This Book

Purpose

This book provides SAS programming professionals with a list of practical ideas that they can use to develop top SAS programming skills and maximize their professional standing in the world of information technology specialists. Using the proven concepts and methods presented in this book, readers can become significantly more proficient SAS programmers, considerably increase their value to their organizations, and appreciably advance their careers.

This book also acquaints SAS professionals (and other interested people) with the wide world of SAS programming and the many resources available to them. Readers will learn how to leverage resources such as SAS users groups, SAS virtual communities, SAS conferences, SAS documentation, the SAS website, and others for maximum benefit. By pursuing the strategies and employing the techniques in this text, readers can maximize their professional standing, earning potential, and job satisfaction.

Is This Book for You?

Novice and intermediate SAS professionals will find this book to be an effective guide to building their SAS programming careers. Advanced SAS programmers will find it useful as a tool for revitalizing their love for SAS programming and for effectively harnessing all of the SAS resources available to them. This book is right for SAS professionals who want to be the top SAS programmer in their organizations, well-known top SAS programmers, top-earning SAS programmers, or all three.

Prerequisites

There are no software prerequisites for taking advantage of the proven methods for becoming a top SAS programmer provided in this book. It is designed for people who are thinking about using SAS software, those who are beginning their SAS programming careers, and those who are experienced SAS programmers. Readers can have experience in any industry or government sector. However, people who are self-motivated and who are willing to learn more and do more than just what their jobs require will benefit the most.

Additional Resources

SAS offers you a rich variety of resources to help build your SAS skills and explore and apply the full power of SAS software. Whether you are in a professional or academic setting, we have learning products that can help you maximize your investment in SAS.

Bookstore	http://support.sas.com/bookstore/
Training	http://support.sas.com/training/
Certification	http://support.sas.com/certify/
SAS Global Academic Program	http://support.sas.com/learn/ap/
SAS OnDemand	http://support.sas.com/learn/ondemand/

Knowledge Base	http://support.sas.com/resources/
Support	http://support.sas.com/techsup/
Training and Bookstore	http://support.sas.com/learn/
Community	http://support.sas.com/community/

Keep in Touch

We look forward to hearing from you. We invite questions, comments, and concerns. If you want to contact us about a specific book, please include the book title in your correspondence.

To Contact the Author through SAS Press

By e-mail: saspress@sas.com

Via the web: http://support.sas.com/author_feedback

SAS Books

For a complete list of books available through SAS, visit http://support.sas.com/bookstore.

Phone: 1-800-727-3228

Fax: 1-919-677-8166

E-mail: sasbook@sas.com

SAS Book Report

Receive up-to-date information about all new SAS publications via e-mail by subscribing to the SAS Book Report monthly eNewsletter. Visit http://support.sas.com/sbr.

Acknowledgments

If the title of this book had been *How to Become a Top Book Publication Team*, then it would have had to include biographies and interviews of the talented SAS staff who worked on this book. I am indebted to the following people for their great professional guidance and judgment in creating a polished book from my original manuscript. My reviewers were Rick Bell, Cindy Cragin, Paul Grant, Chris Hemedinger, Lisa Horwitz, Rick Langston, Julie Petlick, Michael Smith, and Katie Strange. The SAS Press staff was Candy Farrell, Julie Platt, Cindy Puryear, Tate Renner, and Stacy Suggs.

I owe a special thanks to my SAS Press editor Stephenie Joyner for her encouragement, support, and infectious upbeat attitude.

Finally, I would like to thank SAS Acquisitions Editor Shelley Sessoms for believing in this book project and helping me pitch it to SAS Press.

This book is dedicated to the people who have provided me with opportunities for growth throughout my career in information technology. Their support made a difference at key points in my professional life, and I will always be grateful for their kindness: Rose Mooreman, Bill Smith, Jack Costello, Linda Malagisi, Jim Rinaldi, Moe Vaziri, Bob Fake, Richard Anderson, Jim Smith, Mike Rhoads, Duke Owen, Richard Corson, Andrea Glaser, Stan Sachar, Bill Luckey, Marsha Hasson, Debby Vivari, Rick Dulaney, and Karen Tourangeau.

Chapter 1: Become a Top SAS Programmer

Introduction

Congratulations on making the decision to become a top SAS programmer! You have made a significant decision to embark on a course of action that will maximize your professional standing in the world of information technology specialists. To rise through the ranks of other programmers and become a top SAS programmer is not a simple task. It will take careful planning, as well as plenty of time and effort on your part. You have a lot of work ahead of you, but the professional, financial, and personal rewards of being a top SAS programmer will make it all worthwhile. So, let's get started!

Where you go from here depends on where you are starting from. Perhaps you are currently taking computer programming classes and do not yet have your first programming job. Maybe you have been programming in SAS for a few years and are looking for strategies and techniques that will help you to focus on and enhance your career. Or, possibly you have been programming in SAS for many years and need some ideas on how to rekindle your creative spark and revitalize your career. No matter where you are now, this book can help you.

Student Studying SAS Programming

If you are a student studying SAS programming, then you have a great career ahead of you! SAS programming professionals are employed in virtually all sectors of the U.S. economy. They can be found in large and small businesses, in pharmaceutical companies, in local and state governments, in the Federal government, in non-profit organizations, in colleges and universities, in research organizations, and in many, many other industries. SAS professionals are always in demand, the work is interesting, and the disciplines SAS is used in are always evolving with new technologies to master and new software to learn how to use. Salaries for beginning and for experienced SAS programmers have been strong for many years, compared to numerous other professions.

This book presents ideas on how you can pilot your SAS programming career right from the start. It equips you with specific strategies and methods that you can employ early on to distinguish yourself from your peers. This text describes particular tasks you can perform to go beyond the requirements of your job and rise to the top of your chosen profession using SAS software.

Additionally, you will find several chapters that are designed to give you a broad overview of the scope of the world of SAS programming. These chapters present information about a variety of SAS resources that are readily available to you, such as documentation, training, virtual communities, certification, and users groups. You can use those chapters to quickly get up to speed on the wide world of SAS programming and the many resources that you can use to become a top SAS programmer.

Programmer with a Few Years of SAS Programming Experience

Now that you have been programming with SAS for a few years, you have a good sense of what it takes to be reasonably successful at your job. You understand who your clients are: your boss, lower management, upper management, users groups within your organization, external clients, and so on. You have become proficient in the SAS programming tools at your disposal. You understand the amount of effort required to maintain a computer job and the intricacies of working with other information technology professionals to successfully complete assignments. You understand your organization's structure and where SAS programmers fit in, and you have a general sense of what is going on in the computer industry at large. Now, you are trying to determine how to ramp up your own game and figure out what it takes to get to the top.

You are in a very good position, because your career is still young. Bad work habits and complacency have not set in. You are still in a career growth mode. Your management expects that your skills and knowledge will continue to grow, and with them, your salary. So, there is some expectation on the part of management that your career will continue to evolve. But, those expectations will turn out to be short-sighted on your management's part because their expectations are limited to what they regard as the *normal* career growth of a SAS professional in your position.

This book provides guidance on how you can start reshaping your SAS programming career to maximize its growth. It shows how you can build upon your current successes by employing techniques and strategies that go beyond what most of your peers are willing to do. By pursuing these strategies and employing the techniques in this text, you can maximize your professional standing, earning potential, and job satisfaction as you move past your colleagues. And, they won't even see it coming.

Programmer with Many Years of SAS Programming Experience

With many years of SAS programming experience behind you, you find that you are in a very comfortable position. You know you have the programming and business acumen to be successful in your career. You have been through any number of waves of technology shifts—in software, in hardware, and in computer business methodologies. You have experienced or witnessed *Agile development, RDBS, proprietary software, open source, OLAP, HOLAP, ROLAP, mainframes, mini-computers, fourth-generation languages, distributed processing, parallel processing, object-oriented programming, ROI, ETL, cloud computing,* and let's not forget all of those *paradigm shifts*. You have participated in many projects and completed countless assignments. Your experience has allowed you to see what works and what does not work, firsthand, on any number of computer application projects. You have enjoyed a normal degree of professional development, upward mobility in your organization, and salary growth. You like what you have been doing for a living for the past years; however, you feel somewhat restless.

This book can save you from workplace complacency by adding new zest and new direction to your occupation as a SAS programming professional. You can use the techniques in this book to reinvigorate and jump-start a "stale" career. Wouldn't it be exciting to get back to the old days when everything was fresh and there were many professional possibilities ahead of you? You can use the tips and techniques in this book to create strategies that will help you reinvent yourself as a top SAS programmer. You can add new energy and excitement to your information technology career, rekindle your creative spark, and revitalize your occupation. This book can help you claim the top SAS programmer spot that is the culmination of all of the years you have invested in a healthy information technology career.

What Is a SAS Programmer?

So, what exactly is a *SAS programmer*? That question is pertinent because there is some disagreement about the scope, role, and appropriateness of the term *programmer* among information technology professionals. Some use the terms *programmer*, *developer*, and *software engineer* interchangeably. Others draw a distinction between these terms, believing that *developers* and *software engineers* possess additional software development skills that programmers do not have. Some believe that a *programmer* is simply a person who writes original programs and fixes bugs in them without performing more sophisticated requirements, development, and analysis tasks. To them, the term *programmer* is at least a put-down, and at most a pejorative. Not so to us!

The vastness of SAS software offerings also makes the definition of the term *programmer* somewhat problematic. SAS encompasses a wide variety of software products ranging from Foundation Tools (Base SAS, SAS/ACCESS, SAS/CONNECT, SAS Enterprise Guide, and so on) to SAS Analytics (Predictive Analytics and Data Mining, Data Visualization, Forecasting, Operations Research, and so on) to SAS Solutions (Business Intelligence, Data Management, IT Management, and so on). Many of the SAS products and solutions provide powerful GUIs as a means for building and running applications. This book is not targeted toward computer professionals who only use SAS software powered by GUIs. Here, *programmer* refers to those who actually write lines of code using Foundation SAS software such as Base SAS, SAS/GRAPH, SAS/CONNECT, SAS/STAT, SAS/ACCESS, and so on.

This book uses the term *SAS programmer* as shorthand to describe a wide variety of programming professionals who write computer programs using Foundation SAS software. If you are a programmer, a developer, a software engineer, a computer scientist, a statistician, or a systems analyst who primarily codes in SAS software to perform your job, then you are being addressed when this text uses *SAS programmer*. If you are a programmer working in health care, government, research, pharmaceuticals, banking, or insurance who primarily programs in SAS software, then you are a *SAS programmer*. Consequently, the term *SAS programmer* is a simplification for ease of writing. Think how cumbersome this book would be if the title were *How to Become a Top SAS Programmer, Developer, Software Engineer, Computer Scientist, Statistician, or Systems Analyst*!

What Is a Top SAS Programmer?

As you can imagine, the phrase *top SAS programmer* could have many meanings. It could mean that you are the top SAS programmer in your organization. It could mean that you are well-known in the field of computer programming for your expertise in using SAS for a particular discipline. It could mean that you earn top dollars for your SAS proficiency and programming work. With any luck you might be a top SAS programmer in all three of these areas. Let's take a look at each of them.

Top SAS Programmer in Your Organization

Any organization with a population of programmers is going to have a handful of individuals who are considered to be the top programmers. They likely have a top database administrator, a top web programmer, a top SQL programmer, a top SAS programmer, and so on. Such individuals are very knowledgeable about the particular software they have mastered, very experienced in using it for any number of projects, and have proven track records of completing assignments accurately and on time. They usually have the highest salaries possible for programmers in the organization, and usually receive perks such as being sent to training or attending computer conferences. Upper management respects their professional judgment and relies on them to get the work done. Other programmers look up to them professionally, and rely on them for mentoring and for leadership on programming projects.

Perhaps you like the organization that you are currently working for. It provides a professional, respectful work environment with ample room for career growth. The work you do is interesting, and there are a variety of other projects that you could possibly work on that would also be of interest to you. The salaries for SAS professionals are generous for the area you live in, or for workers in the organization, or for the computer industry as a whole. The company offers in-house and off-site computer training, supports your involvement in professional organizations, supports certifications, and sends staff to computer conferences. The company benefits are attractive and maybe you are already vested in the corporate 401K plan, profit sharing, employee stock ownership plan, or have a generous employee leave plan. Maybe there are other environmental factors such as an easy commute, free parking, on-site day care, a subsidized cafeteria, tuition reimbursement, and so on. Whatever the case, you see yourself as being a part of your organization for the long term.

Determining your chances of becoming the top SAS programmer in your organization requires an honest assessment of its possibility. Smaller organizations present more of a challenge for a person who is working on becoming the top SAS programmer. People generally tend to stay in the same job for years, so you could find your way blocked by "old timers." There might not be an abundance of diverse projects that could allow you to exercise and increase your programming acumen to the point where you can demonstrate that you are a top SAS programmer. There might be limited training budgets, limited management support for certification and conferences, and even limited upper management recognition of the value of the IT staff's contributions to the organization. Consequently, your task of becoming a top SAS programmer could be tougher in a smaller organization.

Large organizations usually provide more opportunities for upward mobility. This is true because there are more overall SAS programming positions available in a large organization. There is often a greater amount of turnover among computer professionals, which produces opportunities for people to take their place. A growing organization can also offer more opportunities for upward mobility. As new projects are initiated, staff are either shifted to them or recruited. Management will tend to favor moving high-accomplishing SAS programmers up the chain of responsibility into the new positions on those projects because management knows they are reliable.

To become the top SAS programmer in your organization requires you to understand some of the on-the-ground realities of the business environment you are working in. Here are some of the things you need to identify.

Know which projects or systems are the most important. Some projects and systems are more important to an organization than others in terms of their value to the enterprise. They might be long-established projects such as Corporate Accounting, vital projects such as Payroll, or exciting new initiatives such as Internet data collection systems. Programmers who work on those projects or systems usually have greater opportunities for advancement and visibility than those who do not. Learn who the key players (managers, users, administrators, and so on) and SAS programmers are on those projects. Determine how you can get on board one of the important systems or hot projects. Doing so could increase your chances to contribute to your organization, to participate in programming the latest business strategies, and to increase your visibility, experience, and opportunities for advancement.

Know the standing of other SAS programmers in your organization. Whether your organization is big or small, it likely has other SAS programmers working in it. You should evaluate where you stand in relation to the other SAS programmers in regard to opportunities and promotions. Some of them might be junior to you, some might be your peers, and others might be more senior than you. Junior staff and peers are often easier to pass by with promotions and key assignments as you work at becoming a top SAS programmer. But senior staff might be more problematic because they often have the more valuable and more visible programming assignments, and usually have a good track record of accomplishments. Don't be put off by the fact that such senior staff are enjoying the visibility that they have so rightfully earned. By constantly striving to become innovative, professional, and adept at solving your enterprise's business problems using SAS software, you can rise to the position of top SAS programmer. The techniques in this book and your own ambition can get you there.

The characteristics for being a top SAS programmer in an organization fall into three main categories. You must be:

1. **An expert SAS programmer** – Be adept at using SAS programming language tools. Have a firm command of DATA step programming, SAS procedures, SAS macro programming, PROC SQL, and so on. Understand which SAS tools are the most effective for a particular programming problem.
2. **Expert at completing assignments** – Know the intricacies of your organization's data. Understand your clients' needs. Know how to use corporate resources, from in-house subject matter experts to the computer facilities available to you. Get your tasks done on time and make sure that your reports and result sets conform to the requirements.
3. **The go-to person for SAS knowledge and information** – Be familiar with the most recent versions of SAS software. Know the latest SAS programming language features and constructs. Understand the latest SAS programming techniques. Be a mentor to other SAS programmers in your organization.

Why shouldn't *you* be the person who gets the most challenging assignments, who contributes the most to key projects, who gets the top pay, who is sent to training, who management counts on, and who junior programmers look up to? The answer to all of these rhetorical questions is that there is no good reason to the contrary. You *should* be the top SAS programmer in your organization!

Top as in "Well-Known" in the SAS Programming World

The global SAS programming community has grown larger, year by year, since SAS was founded more than 35 years ago. SAS currently has more than 13,000 employees located in 400 local and regional offices in 55 countries. SAS is used at more than 60,000 sites in 135 countries, including 90 of the top 100 companies on the 2012 FORTUNE Global 500® list. The worldwide SAS community is linked via discussion groups such as the forums on www.support.sas.com, by the sasCommunity.org wiki, by webinars on the SAS website, by SAS newsletters, and by annual users group conferences such as SAS Global Forum and PharmaSUG. With SAS being used in so many organizations by so many computer professionals, it is not surprising that some have emerged as well-known experts in the SAS programming world.

The most well-known SAS programmers are well known because they assert themselves both in writing and in live presentations. They write books for SAS Press, papers for SAS conferences, blogs related to SAS, articles for the sasCommunity.org wiki, and responses in the discussion forums on www.support.sas.com. They present instructional webinars, teach SAS seminars and classes, and present technical papers at SAS conferences. Because they are active in sharing their SAS expertise, their particular names are known over the names of the thousands of other SAS programmers who do great work, but are less vocal about it.

Well-known SAS programmers tend to fall into two general categories:

Recognized experts in a particular area of SAS programming – These are the people that others turn to for information and advice on using a particular feature of the SAS programming language. They are people who have published papers on SAS topics, written extensively in online forums about a particular topic, given presentations about a specific topic, or taught classes based on a distinct topic[1]. Their publications are usually in-depth, easy-to-understand, and thorough. They extend the information already available in SAS technical documentation by applying it to real-life programming situations and providing plenty of examples. Recognized experts often come from the SAS user community, but they can also be SAS staff members. Though their areas of expertise might differ, what they all have in common is that they are passionate about the particular topic that they write about. They know their topic areas in-depth and consequently can explain the details in plain language that anybody can understand.

Such in-depth knowledge comes from extensively using a particular feature of SAS programming. It comes from researching that feature and learning all of its nuances. If it is a SAS procedure, then you need to learn all of the available options of that procedure, the input files, output files, and the various statements available for that procedure. You need to learn the best way to use that procedure and what, exactly, it provides for a given data analysis. You should know something about every facet of that procedure, and you should be able to answer any question about that procedure.

There is no doubt that you can master a particular area of SAS programming. Unless you are just starting out, you probably have a lot of experience with any number of SAS products, procedures, DATA step programming statements, options, functions, macros, and so on. Is there one particular area that interests you, that you have a talent for, or that you have more experience in than others? Perhaps it is not simply one SAS feature, but a way that you use SAS to your organization's benefit, such as data cleaning.

Whatever the case, you can focus on a particular facet, feature, or product of the SAS software suite and become an expert in it. Read the relevant SAS documentation. Read all of the applicable SAS conference papers. Use that programming feature extensively in your own programs, so that you understand its nuances and the best way to leverage it to get the most out of it for your organization. Then, determine how you can add to the body of knowledge of that particular feature of SAS software by authoring your own conference papers, taking part in online forums to offer advice and guidance, or even writing a book about it.

Recognized experts as SAS programming generalists – These are people who are well-known because they have written about or given presentations on diverse SAS programming subjects. Consequently, they are well-known SAS generalists. This means that they have an expert understanding of a wide variety of SAS topics and do not necessarily specialize in a particular topic area. By doing this, they have provided other SAS users with material they can use to learn features of the SAS programming suite. Not only has this helped fellow SAS programmers, but it has raised the visibility of these particular experts and made them *household names* in the world of SAS programming.

The key to becoming an expert SAS programming generalist is to learn as much as you can about as many features of the SAS programming suite and use those features in real-life work situations to process data. Doing so, you gain an in-depth knowledge of SAS programming and are in a position to share that knowledge with other people. Then, you can write your own SAS conference papers and books, and teach SAS classes so that you become a well-known SAS programming generalist.

Top as in Earning Top Dollar for Your SAS Programming Expertise

Another type of *top SAS programmer* is somebody who earns top dollar for their SAS programming expertise. This means that the person earns substantially more than other SAS programmers in an organization or in a given geographic area. Such a person does so because of that person's expertise, experience, or a combination of both. SAS programmers who are well-known are often able to parlay their well-deserved reputations into maximizing their income.

The considerations for earning top dollar for your SAS programming expertise depend on whether you work for an organization or whether you are an independent consultant. Each situation has its own group of factors that can have a direct effect on the amount of money you can reasonably expect to earn, and so must be considered separately.

Working for an Organization

The truth is that any organization will pay only so much to its computer programming staff. Salaries are governed by established organization pay scales, by industry standards, and by geographically defined local pay levels. Many organizations have a *grade level* system for paying staff members. For example, computer professionals might occupy grades 10 – 15, where 10 is at the lower end and 15 is at the upper end that you aspire to. Salaries in grade levels often have a higher and lower boundary. For example, salaries could look like this:

Grade	Title	Salary Range
10	Programmer I	$30k - $40k
11	Programmer II	$35k - $50k
12	Senior Programmer	$45k - $60k
13	Programmer/Analyst I	$55k - $70k
14	Senior Programmer/Analyst	$65k - $80k
15	Programming Manager	$80k - $95k

Within this particular system, you could be hired as a *Grade 11 Programmer II* because of the current level of your SAS programming skills and job experience. It might be expected that you would work in that grade for two or three years as your contributions to the organization became more valuable, your programming expertise grew, and your salary level increased. Then, you would be promoted to a *Grade 12 Senior Programmer* and begin the climb to the next grade level.

This normal career growth cycle can be affected by such factors as the department you work in, the projects you work on, the assignments you are exposed to, the years of SAS programming experience you have, and the depth of experience of the other SAS programmers you are working with.

Learn the grade levels and salaries for computer programmers in your organization. That should not be hard, because many organizations either publish such information on their intranets or have it available in the personnel office. Find out what the requirements are for each level above the one you are currently within. Identify which projects have the greatest possibilities of supporting staff members at the higher grade levels. Maybe you will need to transfer to another group or division to maximize your earning power. You might need to transfer to another project or take on a different role in the project you are currently working on. Whatever the case, you will have to take action to effect a change for yourself.

Assess how realistic it is for you to advance to the next grade level to grow your salary and your expertise. This might necessitate a frank discussion with your manager on what skills and accomplishments you need to have in order to be promoted. Do not be timid; have those discussions, if necessary. Then, work hard to get yourself promoted by using as many of the techniques found in this book as is possible.

Independent Consultant

Independent consultants usually make more money than salaried employees. Sometimes they make a lot more money! But, it comes at a price. Independent consultants have less job stability. They are hired by various organizations for fixed periods of time and can be let go at the convenience of the contracting organization. They are responsible for things that salaried employees take for granted such as paying for their own insurance, health care, social security taxes, Medicare taxes, travel expenses, and paying quarterly estimated income taxes. They always have to be on the lookout for the next consulting opportunity and secure it before the current one ends. They often have to do their own marketing and networking, and might not end up working 50 forty-hour weeks in a year. If they work through a broker, then they have to pay a portion of their hourly billing rate to that broker.

With all of these perceived negatives, you might wonder why anybody would want to be an independent consultant. However, the flip side is that they usually make great money, do not have to deal with the same office politics as regular employees, are exposed to many more applications and computer systems, have a wide variety of assignments, can move on if they encounter an unreasonable boss, and can realize some unique tax benefits. All of this makes consulting attractive to some SAS programming professionals.

The amount of money you can make as an independent consultant depends on a lot of factors. Some of them are the geographic area you are in, the pay scale of the industry, the length of the assignment, the amount of experience and expertise you have, and the type of SAS programming expertise that you have. In some areas, consulting companies have exclusive arrangements with major employers. Consequently, all consultants must subcontract with those consulting companies in order to work at certain organizations. The consulting companies usually either take a cut of the hourly wage or set a specific hourly rate that they will pay you—say a maximum of $50 per hour. It is often difficult or impossible to work around these organizations and get your own terms.

To make top dollar as an independent SAS consultant takes significant research and planning. Here are some of the main steps:

1. **Determine the hourly consulting rates in your area.** Research what consultants are making for SAS programming in your area. You can do this by contacting consulting agencies, researching consulting opportunities in online services, looking at what local SAS programmers are being paid as specified in advertisements on job boards and extrapolating an hourly rate, and by talking to other SAS consultants. Consulting rates are often affected by the overall salary structure of a given geographic area, the industry, the availability of qualified SAS consultants, the particular industry expertise needed for the assignment, the length of the assignment, and the particular SAS skills needed for the assignment.
2. **Consider whether you need to learn more about SAS.** No matter what your level of SAS expertise is, you cannot know everything about SAS. Having additional SAS software skills can make you more valuable. Though Base SAS programmers always seem to be in demand, sometimes specific SAS modules or expertise becomes "hot" and therefore organizations are willing to pay more for consultants with that knowledge. You might consider taking SAS e-learning courses, earning a SAS certification credential, or reading up on and experimenting with a particular SAS module that you haven't already mastered. Perhaps, you could take an initial assignment using a "hot" SAS module, say SAS BI Server, for a lower rate to gain the experience. Thereafter, you could charge a higher rate because you would have gained the expertise.
3. **Decide whether you need a broker.** As previously mentioned, some consulting organizations have contracts with local organizations to provide SAS consultants. They have already done the leg work of finding the opportunity. Working with them, you might be able to come to a good financial arrangement that allows you to maximize your rate and have a good consulting gig. That will call for frank and aggressive negotiations on your part. Knowing the prevailing consulting rates of the area, industry, and particular SAS expertise needed for the assignment is vital to your negotiations. Make sure you do your homework so that you can earn top dollar.Consultants who work through brokers fall into two categories: W2 and 1099 consultants, so named because of the two different IRS documents that they must file. W2 contract programmers work directly for the consulting company as quasi-employees. The company might or might not offer them benefits, but usually collects and pays state, federal, and social security taxes from their earnings. A 1099 consultant is totally independent. Though the consulting company has provided the consulting opportunity, the 1099 consultant is responsible for her own benefits and filing income taxes. Whether you are a W2 or 1099 consultant, the consulting company usually collects money for your work from the client, takes its percentage, and then pays you the agreed-upon hourly fee.

4. **Look for consulting engagements.** Once you have done your homework on the maximum rate that you can expect to earn, you can look for consulting opportunities. If you are going through a broker, then the next steps are largely up to the broker. If you are truly independent, then you need to comb through online job boards, follow up on Internet advertisements for permanent employees to see if they will hire a consultant instead, and network with acquaintances and contacts you have in the business world.There are usually multiple consulting organizations operating in a given geographic area, especially if it is around a major city. Take advantage of this situation by contacting several consulting organizations. Let them know that you are in discussions with other organizations and that you would like for them to give you a better hourly rate. Let them know that you are ready to establish a long-term relationship with them that will maximize both their and your own earning potential. Such negotiations can help you raise your hourly rate and consequently your bottom line.
5. **Determine whether you need to travel.** Sometimes the only way to maximize your consulting rate is to travel to an area where the rate is higher. If you are willing to travel, then you need to determine whether the cost of travel and lodging will still make the higher rate worthwhile. If so, contact principal organizations and consulting organizations in the geographic areas that you believe can maximize your hourly rate. Ask them if they will be flexible enough to let you work from home part of the time.

How Much Do SAS Programmers Get Paid?

So, how much do SAS programmers really get paid, anyway? The answer to that question is the same as the answer to many of life's big questions: *It all depends!*

Whether you are a full-time employee or a consultant, the most basic step in earning top dollar for your SAS expertise is to determine what you are worth. You need to find out what programmers like you make and what the top salary or consulting rate is for a given region, industry, and organization. You can do this by combing through salary surveys, online job boards, discussion forums, newspaper advertisements, and the published salary scales of individual organizations. Once you establish the average salary and the ceiling for SAS programmers with your expertise and experience in a given organization and profession, you can make informed decisions on taking the next steps to maximize your income and earn top dollar as a SAS programmer.

As previously discussed, there are many factors that affect your earning potential as a SAS programmer, so there is no easy answer to the question about how much SAS programmers get paid. Your compensation will be dependent on a plethora of considerations that affect how programmers with any type of software expertise are paid. Here are some of the major factors:

- **An organization's pay scale** – Formal grade levels limit what an organization is willing to pay to employees and to consultants.
- **Your educational background** – A person with a PhD often makes more money than a person with a Bachelor's degree. A person with a Computer Science degree might make more money than somebody with an English degree. Some educational backgrounds pay more because they required more education or they are more applicable to the type of computer work being done in an organization.

- **The industry you work in** – Some industries tend to pay more than others. For example, pharmaceuticals and telecoms tend to pay more than educational institutions.
- **Your job experience** – Factors such as your overall years of programming experience, time spent working in a given industry, and experience with specific SAS products, can make a difference.
- **Consulting agreement** – Whether you are a W2 or a 1099 consultant and the hourly billing rate you are able to negotiate are factors for consultants.
- **Geography** – Enterprises centered around major cities tend to pay more than those in rural settings. There are regional differences as well. The pay scales in the Northeast states are generally higher than those in the Southeast, Central, and Western states.
- **Special considerations** – Security clearance, U.S. Citizenship, PhD in Bioinformatics, and so on, are some of the special considerations that might make a difference in pay and access to jobs for some organizations.

The good news is that SAS programmers receive compensation commensurate with that of other programmers in the information technology industry. You can see this in salary surveys where there is a category for "SAS Programmer." In most surveys or publications, SAS programmers are included in major job function titles such as Programmer or Programmer/Analyst. Here are a few major categories and the average pay for them as of this writing:

- Computer Programmer – $74,000
- Programmer/Analyst – $76,000
- Systems Analyst – $75,000
- Senior Systems Analyst – $85,000

SAS programmers are undoubtedly found among the ranks of those common positions, so they are earning at least those average salaries. And, top SAS programmers are earning more than that.

This is a list of sources you can use to help determine salary information and consulting rates:

- Computerworld Salary Survey
    - http://www.computerworld.com
- Bureau of Labor Statistics
    - http://www.bls.gov
- Monster.com
    - www.monster.com

- Dice.com
  - www.dice.com
- Government Service (GS) pay scale:
  - http://www.fedjobs.com/pay/pay.html

When you have completed your research, you will know whether you are earning top money for your SAS expertise. If not, continue reading this book to glean the techniques that can lead you to a top salary. If so, lend this book to a friend.

A Call to Action!

By purchasing this book, you have made the decision to become a top SAS programmer. The rest of this text provides clear-cut methodologies and strategies designed to get you to the top. These are tried-and-true methods, ideas, and resources. They are things the author has used and done, and things that have worked for many other people who are also top SAS programmers. It does not matter where you are starting from in your career, you can do it. Really! It only requires that you employ your inner strength and determination to set yourself on the course to becoming a top SAS programmer.

This book is your secret weapon for maximizing your SAS programming career. Do not be shy or hesitant in becoming the top SAS programmer in your organization. Sometimes you are the expert SAS programmer just because you say you are. You studied the material, learned the techniques, and performed the work; now you are the resident expert in *all things SAS* in your enterprise. It can be as easy as that. And, why shouldn't it be? Somebody is going to be the top SAS programmer in your organization, in your industry, in your area. Why shouldn't it be you? Well, it *is* going to be you.

So, take a long, deep breath and let it out very slowly. You are now on your way to joining the elite group of top SAS programmers. You will either be the top SAS programmer in your organization, a well-known top SAS programmer, a top-earning SAS programmer, or all three. Whatever, the case, your career will never be the same as you kick it into high gear and take your rightful place at the top. What are you waiting for? Turn the page and let's get going!

---

[1] Topics include such diverse subjects as the Output Delivery System SAS macros SAS hash objects PROC IML, and PROC REPORT.

Chapter 2: Learn Your Craft

Introduction

I would bet that you are among the millions of fellow television viewers who watch talent shows such as *American Idol* or *So You Think You Can Dance*. Part of the fun of the early episodes of these shows is seeing contestants who want to be a top singer or a top dancer but clearly cannot sing or dance. These wannabes appear to believe that with minimal effort and negligible training, they can surpass other contestants who have devoted the time and effort to learning the fundamentals of their art form. They seem to think that just wanting to be a top singer or a top dancer will get them there. Because these aspirants have not put much work into learning the basics of their craft they predictably fail, often with very humorous results.

The same thing is true about becoming a top SAS programmer. Just wanting to become one will not get you there. You need to invest the time and effort it takes to learning the fundamentals of SAS programming. You have to do the work. You have to know how SAS works. You need to learn your craft. Then, you must exercise your knowledge of the fundamentals again and again and again, until you have mastered SAS programming.

With over twenty-four SAS Foundation products available, it can be pretty daunting to know exactly where to begin. SAS Institute Inc. has dozens of free online technical publications, scores of printed technical publications for sale, over 150 SAS Press books for sale, and dozens of SAS classes, all of which provide detailed information about the SAS programming language. There are thousands and thousands of pages online and in print specifying proper SAS syntax and the best ways to use SAS to read data, process data, analyze data, and create result sets. They are your best sources for detailed information about SAS fundamentals.

This chapter offers specific advice on the basics that you should absolutely know to become a top SAS programmer. Should you know more than what this chapter specifies? Of course! This chapter identifies the bare minimum of core SAS essentials that you *must* learn. It is not intended to be a comprehensive list. Nor does it delve very deeply into any particular topic. Instead, it provides brief overviews and checklists of each of the topic areas in order to give you direction and context. Use the checklists as springboards for further research via SAS online documentation, SAS printed documentation, SAS Press publications, or training classes offered by SAS Education.

Once you have identified the components of the SAS language you want to learn (or want to learn more about), see Chapter 5, "SAS Documentation." That chapter contains a description of both online and print references, which contain the detailed explanations that you need. Make sure you bookmark the online references in your browser, and consider obtaining one or more of the print documents for your own personal SAS library.

If you are an experienced SAS programmer, consider having fun with this chapter by determining how many of the topics you know cold. If you are thoroughly familiar with a particular topic, good for you! If you are not acquainted with a particular SAS language construct, or if you have gaps in your knowledge, then do the research and add these items to your personal SAS toolkit. Don't be surprised if there are several subjects that you do not know very well. And don't be surprised if the evolution of the SAS programming language has introduced additional features to old constructs. Simply do the work of filling in the knowledge gaps.

If you are new to SAS programming, use this chapter as a blueprint for further study. Go over the lists and determine which items you do not know. Then, go to the references, read up on them, practice using them, and cross them off your list of unknown topics. But make sure that you don't actually write in this book!

The DATA Step

The DATA step is one of the most important constructs of the SAS programming language because it allows you to write a complete "mini" SAS program from scratch. (This contrasts with SAS procedures, which are "canned" routines that have predefined syntax, formats, options, and parameters). You can write a DATA step to input data other than SAS data, process existing SAS data sets, interact with your operating system, create new SAS data sets and variables, create flat files, and do a host of other exciting things. The DATA step is a wide-open canvas on which you can practice your programming skills using a rich bounty of SAS statements.

Here are constructs that you should understand or know how to use in a DATA step:

ERROR automatic variable is set to 1 when SAS encounters an error in the DATA step.

_N_ automatic variable stores the value of the number of times the DATA step has iterated.

BY statement and the FIRST.*variable* and LAST.*variable* variables are powerful ways to make processing decisions when accessing the first value or the last value for variables that were used to sort the SAS data set.

DATA statement is where it all begins. This statement is used to create one or more SAS data sets.

- _NULL_ option directs SAS to *not* create a permanent data set during execution of the DATA step.
- VIEW= option is used to create a data view file instead of a SAS data set.
- PGM= option specifies the name of the stored compiled program SAS creates or executes in the DATA step.

DATALINES statement specifies that data is in the body of the DATA step.

DELETE statement stops processing the current observation and returns to the DATA statement, effectively *deleting* that observation from the output data set.

Directory handling functions facilitate processing files in a directory from a DATA step.

- DOPEN allocates a directory.
- DNUM identifies the number of files in a directory.
- DREAD makes the name of a file available in the directory.
- DCLOSE deallocates a directory.

DO statements execute blocks of SAS statements and execute statements iteratively based on specified start or stop criteria.

- Simple DO statement specifies a block of statements to be executed in sequence.
- Iterative DO statement executes a group of SAS statements repeatedly. There are five flavors:
    - Start value only
    - Start TO stop
    - Start TO stop with a BY
    - Start TO stop with UNTIL
    - Start TO stop with WHILE
- DO UNTIL statement executes statements *until* a condition becomes true.
- DO WHILE statement executes statements *while* a condition is true.

ERROR statement creates custom error messages that are printed in the SAS log.

File handling functions enable access to file information from within a DATA step.

- FILENAME assigns a fileref to an external file, directory, or output device.
- FOPEN opens an external file via an existing fileref and returns a file identifier valve.

- FREAD inputs a record from the external file and places it in the File Data Buffer.
- FGET gets data from the File Data Buffer and inputs it to a SAS variable.
- FCLOSE closes an external file.
- FOPTNUM, FINFO, and FOPTNAME extract information about the attributes of a file.

FORMAT statement specifies formats for variables.

Hash objects store and manipulate data temporarily stored in computer memory to drastically reduce elapsed time. Understand what hash objects are and when to use them so that you can perform the following tasks:

- declare and instantiate hash objects
- define keys and data
- store and retrieve data in hash objects
- replace and remove data in hash objects
- write hash object data to SAS data sets

IF-THEN/ELSE statement conditionally executes SAS statements based on a value or an expression.

INPUT statement is used as the primary tool to input data stored in files other than SAS files.

LABEL statement creates meaningful labels for variables.

Match-merging merges SAS data sets: one-to-one, one-to-many, and many-to-many.

- MERGE statement directs SAS to join observations from two or more SAS data sets into a single SAS data set.
- BY statement specifies the variables used as "key" variables to facilitate the merge.
- IN= data set option and an IF statement can be used to control which observations are written to the output data set.

OUTPUT statement directs output to multiple SAS data sets in a DATA step.

Program Data Vector (PDV) is an area in memory where SAS builds a new SAS data set one observation at a time.

PUT and PUTLOG statements write information of interest to you to the log.

SET statement allocates one or more SAS data sets for input to the DATA step.

- END= option specifies a temporary variable that acts as an end-of-file indicator. The variable is initialized to zero and is set to one when the last observation of the last input data set is reached.
- KEY= option identifies the name of a simple or composite index used to dynamically read the SAS data set.

- INDSNAME= option creates a variable that stores the name of the SAS data set that the current observation was read from.
- NOBS= option creates a temporary variable that contains the total number of observations in the input data sets.
- POINT= option is used to directly access observations by observation number.

Subsetting IF statement excludes observations based on a value or expression.

WHERE statement and WHERE data set option subset data as it is being input to a DATA step (or a procedure).

SAS Procedures

SAS procedures are special routines pre-written by the brainiacs at SAS to perform particular programming functions. For example, PROC PRINT prints, PROC SORT sorts, PROC MEANS calculates descriptive statistics, PROC COPY copies, PROC GPLOT creates graphical plots, PROC REPORT creates reports, and so on. Some procedures perform utility functions such as copying and sorting, while others invoke complex algorithms that calculate sophisticated statistics. The benefit of using SAS procedures is that all of the behind-the-scenes work for setting up the utilities or performing the calculations has already been done for you.

Procedures have a series of predefined statements, options, and keywords which you can choose from to elicit the desired computation or the function of the particular SAS procedure you are executing. You simply plug in the correct data, options, and statements and then let these canned routines perform the heavy lifting. All of the *parameters* for SAS procedures are well documented in SAS publications, so those are your best guides to using them. Choose the correct procedure, and then plug and play. What could be easier?

Here are the basic SAS procedures that you should know:

CONTENTS procedure creates a *data dictionary* type listing of the contents of SAS libraries and SAS data sets.

- DIRECTORY option prints a list of the SAS files in the library.
- DETAILS option includes the number of observations, variables, indexes, and the data set labels in the output.
- MEMTYPE option restricts output to information for a particular type of SAS file (for example, DATA or VIEW).
- NODS option directs SAS to not print information for the individual SAS files.
- ORDER option prints variables in a specific order, such as alphabetic order or the order in which they occur in the SAS data set.
- OUT option creates a SAS data set containing library, data set, and variable information.

COPY procedure copies and moves SAS files.

- EXCLUDE statement excludes files or file types.
- MEMTYPE option restricts processing to one member type such as DATA or CATALOG.
- MOVE option moves files instead of copying them.
- SELECT statement selects files or file types.

DATASETS procedure provides a plethora of utilities for managing SAS data sets. Know how to use it to perform SAS file management tasks.

- CHANGE statement renames SAS data sets, catalogs, views, and so on.
- DELETE statement deletes SAS files.
- MODIFY statement can be used to change set data set attributes.
- FORMAT, INFORMAT, LABEL, and RENAME statements can be used to change variable metadata.
- LABEL statement creates data set labels.
- GENNUM and GENMAX options control generation data sets.
- INDEX statements creates SAS indexes.
- AUDIT statement controls audit trails.
- IC statements are used to manage integrity constraints.
- ALTER, READ, WRITE, and PW options create, modify, and delete SAS data set passwords.

FORMAT procedure creates custom formats for displaying variables in SAS data sets.

- VALUE statement is used to create a custom format for SAS variables.
- CNTLIN options specifies a SAS data set that will be used to create the format.
- FMTLIB option prints information about existing formats and informats.

FREQ procedure creates frequency counts of the values found in specified variables within SAS data sets.

- BY statement separates analyses for each BY group.
- EXACT statement performs exact tests.
- OUTPUT statement is used to store output into SAS data sets.
- TABLES statement specifies tables.
- TEST statement tests for measures of association and agreement.
- WEIGHT statement specifies weight variables.

MEANS/SUMMARY procedures compute descriptive and summary statistics. These two procedures perform most of the same functions.

- BY statement produces separate statistics for each BY group
- CLASS statement defines the subgroup combinations for the analysis
- ID statement includes variables in the output data set.
- TYPES statement is used to distinguish combinations of class variables.
- VAR statement identifies the analysis variables.
- OUTPUT statement stores the output into SAS data sets.

OPTIONS procedure prints the current values of SAS options in the log

- GROUP option displays the options in one or more groups.
- LISTGROUPS option displays the available option groups.
- OPTION option displays a single option's setting.
- VALUE option prints options, values, scopes, and how they were set.

PRINT procedure prints custom reports.

- BY statement specifies variables to create BY groups.
- ID statement is used with the BY statement to create better looking reports.
- PAGEBY statement starts a new page when the specified variable values change.
- SUM statement creates summary totals of the specified numeric variables in the report.
- VAR statement specifies which variables are to be printed.

PRINTTO procedure specifies an alternative destination for SAS log and SAS procedure output.

- LOG option identifies where to send SAS log output.
- PRINT option specifies where to send SAS procedure output.
- NEW option replaces the contents of an existing file with new output from the SAS log or procedure.

REPORT procedure enables you to create world-class reports from your SAS data sets. Make sure you master at least the basics of this powerful procedure.

- BREAK statement prints a default summary at the change in value for a group.
- BY statement creates a separate report for each BY group.
- COLUMN statement specifies the arrangement of columns and headings that span columns.
- COMPUTE/ENDCOMP statements define compute blocks in which you define how new values can be computed from existing variables.
- DEFINE statement specifies how to display report items.

SETINIT procedure prints a list of the SAS products you currently license to the log.

SORT procedure sorts data into a specified order.

- BY statement lists variables to sort by, including the ASCENDING and DESCENDING options.
- NODUPKEY option removes duplicate records based on duplicate sort key values.
- NODUPREC option removes duplicate records based on all variables being duplicate.
- DUPOUT option creates a data set of duplicate observations when using the NODUPKEY or NODUPREC option.
- NOUNIQUEKEY option removes from the data set observations that have unique key variable values.
- OUT option specifies an output data set.
- UNIQUEOUT option creates a data set of unique observations when used with the NOUNIQUEKEY option.
- TAGSORT option facilitates a sort using a minimum of disk space. This is a performance option.

SQL procedure allows you to use Structured Query Language (SQL) concepts and syntax to process SAS data sets. Consequently, you think of SAS data sets as *tables*, observations as *rows*, and variables as *columns*. Here are the basics you should know:

- PROC SQL statement initiates a SQL query and specifies options that will be used during the execution of the query.
- CONNECT statement is used to create a connection with a database via SAS/ACCESS software.
- CREATE TABLE statement creates a new table.
- DROP statement deletes tables, views and indexes.
- GROUP BY clause is used to group data by specific columns and produces summaries.
- INSERT statement adds rows to a table.
- ORDER BY clause sorts rows into a specified order.
- SELECT statement selects rows and columns from tables.
- UPDATE statement modifies columns values.

TABULATE procedure creates tabular reports from SAS data sets.

- BY statement causes each BY group to be printed in a separate table on a separate page.
- CLASS statement identifies the variables used to create categories.
- TABLE statement describes the specifics of how tables should be created.
- VAR statement identifies the analysis variables.
- WEIGHT statement specifies the weight variables used in statistical calculations for the analysis variables.

TRANSPOSE procedure pivots the data in a SAS data set by turning variables into observations.

- BY statement identifies the variables to use for BY groups.
- COPY statement specifies which variables should be moved without being transposed.
- ID statement lists variables whose values name the transposed variables.
- PREFIX= / SUFFIX= options enable you to specify a prefix or suffix that will be used to name transposed variables.
- VAR statement identifies variables to be transposed.

SAS Performance

Once you have mastered basic DATA step and procedure programming, you can turn your attention to making your SAS programs leaner and more efficient. Making programs efficient reduces the computer resources they consume when running, resulting in faster program execution times. Reducing the footprints of your SAS programs' computer resources also makes you a better corporate citizen by reducing the load on your enterprise's servers, disk storage, and network bandwidth. So take the time to learn what you can to craft efficient SAS programs. Then, modify your programs, document the computer resources you have saved, and show the results to your manager. She will be suitably impressed!

Performance techniques Become familiar with the following SAS performance techniques:

- Limiting variables and variable sizes to keep your SAS data sets *thinner.*
    - The DROP statement and option drop variables from a SAS data set.
    - The KEEP statement and option keep specified variables in a SAS data set.
    - When possible, store numeric data in less than 8 bytes.
    - Rightsize character variables instead of taking the SAS default sizes.
- Limiting observations will keep your SAS data sets *shorter.*
    - Subsetting IF statements can be used to keep only observations that match a specific condition.
    - The WHERE statement can subset SAS data sets based on existing values within the data sets.
    - Know when to use IF versus WHERE.
- VIEWS – Views are small files that contain instructions on how to process data from SAS data sets. Views are an alternative to creating SAS data sets, which contain observations.
    - Know how to create a view in a DATA step.
    - Understand how to create a view in PROC SQL.
    - Be aware of how views can save disk space.

- Updating SAS data sets in place to reduce input/output operations.
  - Learn the basics of how to append one data set to another with PROC APPEND
  - Discover how you can update a master data set with a transaction data set using the MODIFY statement. Also learn the advanced techniques of checking the status code to determine whether updates have taken place.
- SASFILE statement – Put SAS data sets in memory for quicker read access.
- SAS indexes – Facilitate quick access to small subsets of observations in large SAS data sets.
  - Learn about *simple* and *composite* SAS indexes.
  - Understand how to create and remove SAS indexes.
  - Know how to use SAS indexes in your programs.
- System options that can reduce input/output operations.
  - The BUFNO option specifies the number of buffers to allocate for each SAS data set.
  - The BUFSIZE option specifies the page size of new SAS data sets.
  - The CBUFNO option specifies the number of buffers to allocate for SAS catalogs.
  - The IBUFNO option specifies the number of buffers to allocate for SAS indexes.

Measuring performance You can measure SAS performance using the following:

- The FULLSTIMER option writes performance metrics to the SAS log.
- The Logparse SAS macro captures program performance metrics in a SAS data set.
- The Interface to Application Response Measurement (ARM) can be invoked in SAS programs in order to measure their performance.
- The RTRACE facility can be used to measure SAS usage.

Miscellaneous SAS Programming Elements

There are a number of important SAS programming elements and constructs that are so unique that they could easily form their own categories. Rather than do that, they have been *lumped* together into this section. Don't let that fact diminish their importance. Each of these constructs is something you should know, something that is basic to your better understanding of SAS programming. So take the time to learn them and see if they seem so *miscellaneous* to you.

SAS Functions There are hundreds of SAS functions, so it is impossible to know them all. Simply be familiar with the various categories of functions, where to find information about them, and some of the more often used functions. Some of the categories you should be the most familiar with are:

- Numeric functions
- Character functions
- Datetime functions

- Descriptive statistics functions
- Distance functions
- External file functions

CALL Routines CALL routines are similar to functions, and there are scores of them. Be familiar with some of them.

- Understand what CALL routines are and where to find information about them.
- Some of the more utilitarian CALL routines are CALL SYSTEM, CALL SYMPUT, and CALL SYMGET.

Output Delivery System (ODS) The Output Delivery System is a powerful tool for generating sophisticated reports in many different output formats. Understand the basics:

- creating PDF, RTF, CSV, and HTML output
- specifying ODS destinations
- ODS language statements
- ODS table templates, table elements, and table attributes
- ODS styles, style elements, and style attributes
- item stores, template stores, and directories

SAS Enterprise Guide SAS Enterprise Guide is a sophisticated GUI environment that enables you to build, schedule, and execute SAS programs through a point-and-click interface. It has a very robust programming node with autocomplete and context-sensitive programming Help that can facilitate learning SAS programming constructs. SAS Enterprise Guide comes with Base SAS and is well-worth learning, because it represents SAS' strategic direction for future programming environments.

- create a project in the Process Flow window
- add SAS data sets to your project
- input data sets that aren't SAS data sets to your project
- add SAS tasks to your project to create reports and graphs
- understand how to use the query builder to select, order, and create columns; to select and order rows; to join tables; and to group and summarize data
- learn how to use the New Report window to combine graphs and tables onto the same page
- determine how you can create sophisticated process flows that take raw data and run it through a number of steps to produce the desired output, using SAS tasks and existing SAS programs

Dictionary Tables and SQL Views These are two sources of metadata that you can use in your SAS programs.

- Run-time source of data sets, catalogs, columns, options, macros, and so on
- dictionary.<*table name*> - Know some of the more useful tables:
    - dictionary.tables
    - dictionary.columns
    - dictionary.destinations
- sashelp.<*viewname*> - Know some of the more useful views:
    - sashelp.vextfl
    - sashelp.voption

SASHELP library This library contains some useful SAS files.

- Sample SAS data sets can be used to practice SAS programming techniques.
- Views files contain metadata of SAS files, external files, and other SAS facilities.

SAS Display Manager (DMS) This interactive windowing environment provides facilities to write, save, and execute SAS programs. It is available in SAS for Windows, UNIX, Linux and z/OS.

- DMS automatically allocates five windows:
    - Results window displays the results of SAS programs.
    - Explorer window enables you to explore SAS libraries, flat files, and directories on your computer.
    - Enhanced Editor window features a sophisticated editor for creating and running SAS programs.
    - Log window displays the SAS log of all programs executed during a DMS session.
    - Output window displays reports and other SAS program output you send to the Listing destination.
- *Customize settings* is a facility to customize DMS settings such as what is on the menu bars, enhanced editor options, key settings, and the default folder SAS is to use to open and store programs in.
- *File Import/Export wizards* are GUI utilities that enable you to import various files into SAS data sets and to save SAS data sets in various other data formats such as CSV, TXT, and XLS.
- *Solutions drop-down list* provides a GUI interface to other licensed products such as SAS Enterprise Miner, Market Research, Time Series Forecasting, SAS/EIS, and SAS/ASSIST.

Format catalogs – Understand how format catalogs are allocated and associated with SAS data sets.

- FMTSEARCH option specifies the order in which SAS searches format catalogs for a particular format.
- INSERT and APPEND options enable you to specify a format catalog for SAS to search either before or after other already allocated catalogs.
- NOFMTERR system option directs SAS to ignore the fact that it cannot find a format for a particular variable.

SAS macros – The SAS Macro language is a powerful tool for creating generalized, re-usable SAS programs, and for generating SAS code. There is much to know about SAS macros, most of which is covered in several SAS publications.

You should at least be familiar with the following:

- automatic macro variables
- how to display the values of macro variables to the log
- creating and reading macro variables
    - CALL symput
    - CALL symget
    - INTO clause in PROC SQL
    - Scope of macro variables – Global versus local macro variables
- macro language programming
- macro expressions
- macro quoting
- compiling macros
- debugging macro options
    - MPRINT
    - MLOGIC
    - SYMBOLGEN
- storing and reusing macros
- autocall libraries

SAS and security The three most important facets of your organization's data are security, security, and security! Consequently, you should understand the many SAS security features you can use to protect your organization's SAS data sets.

- Passwords
    - Be familiar with the four types of passwords: ALTER, READ, WRITE, and PW.

    - Know how to create, modify, and delete passwords.
    - Recognize how to specify passwords in DATA steps and procedures.
- Understand how to encrypt SAS data sets.
- The READONLY LIBNAME option protects data sets in a data library from being updated.
- The NOREPLACE system option stops permanent SAS data sets from being updated.
- The PWENCODE procedure is used to encode passwords to keep them secret.
- Audit trails keep track of who updated permanent SAS data sets, when they did it, and what was changed.

SAS autoexec and config files These two system files allow you to configure your SAS environment.

- Understand the purpose of each file and what you can do with it.
    - Autoexec.sas
    - Sasv9.cfg
- Know where these files can be found within your SAS installation.
- Be aware of what you can do with them.
- Be able to specify autoexec and config files in batch programs or the SAS shortcut on the desktop.

Sending commands to your operating system SAS has several facilities for sending commands from a SAS program to the operating system.

- The CALL SYSTEM CALL routine sends a command to the operating system in a DATA step.
- The X command sends a command to the operating system in open code.
- The RENAME function facilitates renaming files and directories from within a DATA step.

Documenting your SAS programs You absolutely must document your SAS programs.

- Two simple ways of creating comments:
    - /* ... */
    - *.... ;
- Put plenty of comments in your programs to document when they were written, the input and output files, and their purpose. Document major DATA and PROC steps so that others can follow the logical flow of the program.
- Put a comment header section at the beginning of a program.
- Document major code blocks.
- Give programs meaningful names.

System options for debugging These options can save you a lot of time when debugging your programs.

- The MSGLEVEL=I option writes informational SAS notes, such as index usage and sort notes, to the log.
- The SOURCE2 option prints the source code for %INCLUDE statements.
- The SYMBOLGEN, MPRINT, MLOGIC options write macro processing code and logic to the log.

Cross Environment Data Access (CEDA) These powerful engines enable SAS to read SAS data sets created on servers with different data representations.

- Understand the concept of CEDA engines
- Know where to find host information
    - Data representation
    - Host created
- Understand the limitations to using SAS data sets created on foreign hosts
- The OUTREP= data set option and LIBNAME option create SAS data sets with a foreign host's data representation

Learn How to Input Data Sources Other Than SAS

It would be wonderful if every bit of information that you had to process was stored in SAS data sets. But in real life, data is stored in a staggering number of file formats that are not SAS. Data is stored in Excel spreadsheets, Microsoft Access databases, comma-separated value files, fixed-length flat files, variable-length flat files, tab-delimited files, XML files, HTML files, VSAM files, Sybase database tables, SQL Server database tables, DB2 database tables, and Oracle database tables to name a few. So, an important part of SAS programming is knowing the proper tools to use to input these data sources into your SAS programs and your SAS data sets. Fortunately, SAS has facilities for reading, writing, and processing data stored in all of the aforementioned data sources as well as many more.

This section describes some of the most common data sources and the SAS facilities for processing them. It provides lists of what you should know to successfully process data in flat files, Excel files, XML files, and relational databases. Make sure that you understand the basics of processing each data source that exists in your own organization. For recommended reading, see the references in Chapter 5 and ramp up your knowledge on how to process these data sources.

Flat files Have a good grasp of the various types of flat files that exist and the tools you can use to input them into your SAS programs. Base SAS has all of the tools you need to process flat files.

- Common types of flat files
    - Comma-Separated Value (CSV)

  - Delimited
  - Non-Delimited
- Processing flat files
  - INFILE and INPUT statements with a DATA step
- INFILE arguments, such as FILEREF, PIPE, and TEMP
- INFILE options, such as BLKSIZE, DELIMITER, EOF, LENGTH, and MISSOVER
- INPUT styles
  - List input
  - Column input
  - Formatted input
  - Named input
- PROC IMPORT syntax
  - Main options – DATAFILE, OUT, and REPLACE
  - Data source statements – DATAROW, DELIMITER, GETNAMES, and GUESSINGROWS
  - The Import Wizard
  - Know how to find the Import Wizard in SAS Display Manager
  - Understand the choices
  - Learn how to save the source code in a SAS data set

Excel files Excel is so ubiquitous that you will undoubtedly end up processing data from Excel spreadsheets sooner or later. Take the time to learn how, now. Some key things to know are:

- SAS/ACCESS to PC File Formats
  - LIBNAME statement
  - PROC IMPORT
  - Import Wizard in SAS Display Manager
- SAS/ACCESS to ODBC
- SAS/ACCESS to OLE DB
- DDE

XML Mapper The Extensible Markup Language (XML) has become a popular standard for sharing data across the Internet and between operating systems. Be aware of SAS' facility for processing XML files (XML Mapper) in case you need to transform XML files into SAS data sets.

- *SAS x.x XML LIBNAME Engine: User's Guide* – The user's guide for the version of SAS that you are currently using.
- http://support.sas.com/documentation/cdl/en/engxml/62845/HTML/default/viewer.htm#titlepage.htm

SAS/ACCESS to Relational Databases If your organization has data stored in relational database, you want to make sure that it has licensed the proper SAS/ACCESS software to enable you to access the data. If so, then become familiar with the specifics of the SAS/ACCESS software that enables you to process the particular databases in your organization. Here are some of the possible SAS/ACCESS products for processing databases:

- SAS/ACCESS Interface to DB2
- SAS/ACCESS Interface to Informix
- SAS/ACCESS Interface to Microsoft SQL
- SAS/ACCESS Interface to MySQL
- SAS/ACCESS Interface to ODBC
- SAS/ACCESS Interface to OLE DB
- SAS/ACCESS Interface to Oracle
- SAS/ACCESS Interface to Sybase
- SAS/ACCESS to SYSTEM 2000
- SAS/ACCESS to R/3

SAS software developers are good at developing ways to access common data sources that emerge from the IT industry. So, if you do not see a particular type of file or database you need to access in the preceding list, check the SAS Products web page. Odds are that there is a SAS tool for processing it there. If you do not find it there, your SAS representative will be glad to help.

Chapter 3: Know Your Environment

Introduction

It is a no-brainer that you know your own home environment. You know which cabinets the pots and pans are stored in, which drawer houses the knives, forks, and spoons, where the pot holders are, which shelves the various sized plates can be found on, and where the drinking glasses and coffee cups are kept. You know where in the refrigerator the meat, cheese, vegetables, milk, juice, and fruit are stored. You know which cabinet drawers and which closets house your work, casual, and formal clothing. So, when you have to cook or eat, or get dressed, your knowledge of your home environment enables you to efficiently move about and use the various resources found in your home to get the task done.

The same is true for using SAS in your work environment. You need to know some very basic things about the SAS environment and the computing environment at work to be effective. A good knowledge of the SAS software you have on hand and of the operating system (or systems) you are using enables you to more easily and effectively leverage SAS to crunch your organization's data in order to surface the information that is required. So, it is important for you to take the time and the effort to know your environment.

This chapter discusses three important facets of knowing your environment. First, you must know which version of SAS you are using and the nuances of that version. Secondly, you should know which SAS products are available in your organization. Finally, you need a basic knowledge of the operating environments that you are using to execute SAS programs. Knowledge of these three central environmental factors will help you to become a more knowledgeable and effective SAS programmer.

Know Which Version of SAS You Are Using

Like a fine wine, SAS keeps getting better and better with age! Every new version of SAS includes additional features designed to increase the utility of the SAS programming language. Many of the new features were introduced as a result of SAS professionals like yourself communicating the need for them via the SASware Ballot. The SASware Ballot is an online survey in which SAS customers can specify and then vote for enhancements to SAS software and services that they would like to have included in SAS. So, many new SAS features come about as a result of real-world professionals identifying a need for them. You can incorporate the fresh features of a new version of SAS into your programs to make them more versatile and more efficient. Doing so keeps you on the cutting edge of SAS programming.

For example, consider some of the new features that were introduced back in SAS 9.2:

- ODS Graphics – This feature enables SAS statistical procedures to automatically create graphs along with the tables they normally produce. In the past, additional programming was required to create graphs from data culled in SAS statistical procedures. ODS Graphics is a significant step forward in producing publication-ready graphical output from SAS statistical procedures.
- Checkpoint Mode and Restart Mode – You can implement your SAS batch programs to use these "modes" so that you can restart programs from the DATA or PROC step in which they failed. When you do, SAS saves all Work data sets so that no intermediate results from previously executed steps are lost if your program has an error during execution. This can save you significant time for long-running production programs that fail deep within the program.
- PROC FCMP – You can use PROC FCMP to create your own SAS functions, CALL routines, and subroutines. So, if you have ever wished that SAS contained a particular function, you do not need to wait for it to be added via the SASware Ballot. Instead, you can use the FCMP procedure to create and store that function. Then, you can use it in all of your SAS programs.
- Interface to R in SAS/IML Studio – The SAS/IML Studio IMLPlus language includes functions you can use to transfer data between SAS data sets and R data frames, and between SAS/IML matrices and R matrices. "R" is a popular statistical programming language and software environment for statistical programming. This interface enables SAS professionals to use SAS/IML Studio to interact with R data and with R statistical procedures.

As you can see, these features that were introduced in SAS 9.2 add new functionality and new utility to SAS, thus increasing the scope of what you can do with it. So, it is important that you keep up-to-date in order to add tools to your SAS skill set that enable you to be a more effective SAS programmer.

Fortunately, SAS makes it easy for you to learn about the new features and services of SAS. They are well-documented in a section of web pages on http://support.sas.com. To find the *What's New* pages:

1. From http://support.sas.com, click the **Knowledge Base** tab at the top of the page.
2. Click **Documentation** in the tree on the left.
3. Click **What's New in SAS** under the Documentation tree.

The *What's New* page provides links to comprehensive documentation that specifies what is new in SAS for several of the last releases. It is your jumping-off point for browsing the latest features and for finding features you might have missed in the previous couple of releases.

Here is a screenshot of the *What's New* page as of this writing.

Figure 3.1: What's New Page

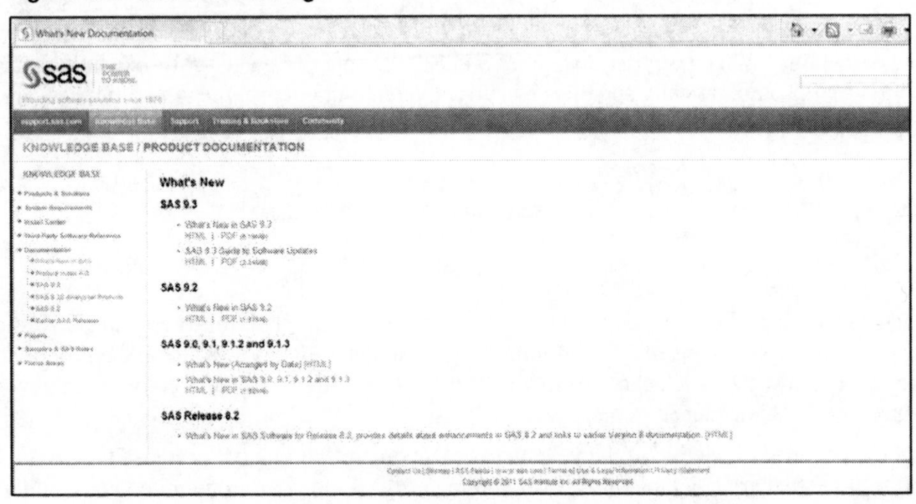

This page provides access to documentation on the latest release of SAS plus five previous releases. There is a lot of good reading in those web pages!

So, what version of SAS do you have in your organization? Do not be disappointed if you cannot answer that question off the top of your head...*this time*. But, it is definitely something you should know. Luckily, it is fairly easy to determine. Here are two quick and easy ways to find out what version of SAS you are running:

- **SAS Display Manager** – In SAS Display Manager, you can determine the current version of SAS by clicking on **Help** at the top menu and then **About SAS 9** from the drop-down menu. You will see an About SAS 9 window that looks something like this:

Figure 3.2: About SAS 9

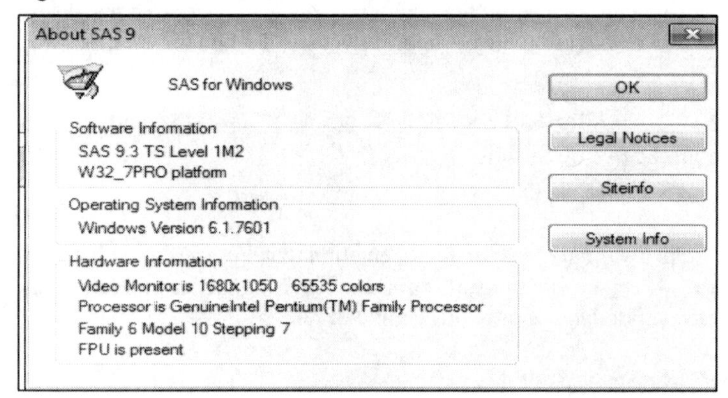

The information in the window tells you the version of SAS as well as information about the maintenance level, operating system, and hardware.

- **The SAS Log** – When you first invoke SAS in SAS Display Manager or by executing a batch program, SAS automatically specifies its current version in notes at the top of the SAS log. It looks something like this:

```
NOTE: Copyright (c) 2002-2010 by SAS Institute Inc., Cary, NC, USA.
NOTE: SAS (r) Proprietary Software 9.3 (TS1M2)
Licensed to MMMMIIIIKKKKEEEE INC, Site 0427.
NOTE: This session is executing on the W32_7PRO platform.
```

Documentation of the SAS version in the log can be especially helpful when you are reading logs from past executions of SAS programs. You can determine what version the programs were executed with, which can inform you of ways they might be improved by using new features of SAS introduced in more recent releases.

Armed with the knowledge of the version of SAS you are running, you can use the products and features that are available in that version to deftly tackle your programming assignments.

Know the SAS Products Available in Your Organization

SAS contains more than 24 SAS Foundation products, and with any luck your organization licenses most or all of them. Because SAS is your primary tool for processing data and surfacing information, you should know exactly which SAS products are available for your use in your organization. If you are running SAS on multiple operating systems, such as Windows, UNIX,

Linux, and z/OS, you need to become aware of which SAS products exist on the various computing platforms. Fortunately, this is an easy thing to do.

There are several ways to determine the SAS Foundation products available to you:

1. PROC SETINIT – This SAS procedure produces a list of information about the duration of your SAS license and a list of the SAS products you have licensed. Here is how it is coded:

   ```
   proc setinit;
   run;
   ```

   The SETINIT procedure produces a message to the log that looks like this:

   ```
   Original site validation data
   Site name:     'MMMMIIIIKKKKEEEE INC'.
   Site number:  04271965.
   Expiration:   30NOV2020.
   Grace Period:  45 days (ending 14JAN2020).
   Warning Period: 45 days (ending 28FEB2020).
   System birthday:   04APR1985.
   Operating System:   W32_WKS .
   Product expiration dates:
   ---Base SAS Software
            30NOV2015
   ---SAS/STAT
            30NOV2015
   ---SAS/GRAPH
            30NOV2015
   ---SAS/FSP
            30NOV2015
   ---SAS/IML
            30NOV2015
   ---SAS/CONNECT
            30NOV2015
   ---SAS/EIS
            30NOV2015
   ---MDDB Server common products
            30NOV2015
   ---SAS Enterprise Guide
            30NOV2015
   ---SAS/ACCESS Interface to Oracle
            30NOV2015
   ---SAS/ACCESS Interface to PC Files
            30NOV2015
   ---SAS/ACCESS Interface to ODBC
            30NOV2015
   ---SAS Workspace Server for Local Access
            30NOV2015
   ```

   Look at all of those great SAS products available for processing and analyzing data!

2. **The ViewRegistry Report** – This handsome HTML report can be produced by executing a Java utility program that is provided by SAS and loaded onto your computer when SAS is first installed. The report looks like this:

Figure 3.3: Example of the SAS Installed Software and Components report.

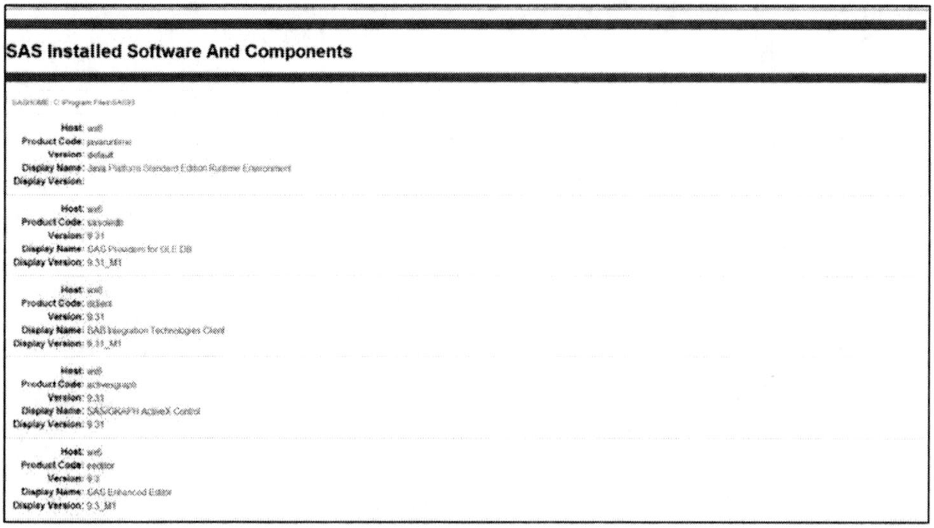

SAS Installed Software And Components

Host: [illegible]
Product Code: [illegible]
Version: default
Display Name: Java Platform Standard Edition Runtime Environment
Display Version:

Host: [illegible]
Product Code: [illegible]
Version: 9.31
Display Name: SAS Providers for OLE DB
Display Version: 9.31_M1

Host: [illegible]
Product Code: [illegible]
Version: 9.31
Display Name: SAS Integration Technologies Client
Display Version: 9.31_M1

Host: [illegible]
Product Code: [illegible]
Version: 9.31
Display Name: SAS/GRAPH ActiveX Control
Display Version: 9.31

Host: [illegible]
Product Code: eeditor
Version: 9.3
Display Name: SAS Enhanced Editor
Display Version: 9.3_M1

The following SAS Usage Note on the SAS support website provides information about where the Java utility program is located on various operating systems:

Usage Note 35968: Using the ViewRegistry Report and other methods to determine the SAS® 9.2 and SAS® 9.3 software releases and hot fixes that are installed

http://support.sas.com/kb/35/968.html

3. **The SAS Installation Reporter Program** – This program provides the most comprehensive set of reports of the SAS products available in your environment as well as a lot of other very good information about your SAS installation. It can be downloaded from http://support.sas.com/kb/20/390.html.

   The SAS Installation Reporter provides information about the following:

   - software your organization has licensed such as Base SAS, SAS/ACCESS, and so on.
   - software your organization has installed — Note that not all of the licensed software might have been installed.
   - SAS clients and applications that are installed such as SAS Enterprise Guide and the SAS System Viewer.
   - current SAS hot fixes that have been applied.

It also provides other interesting SAS installation facts. The output is so long and involved that it cannot fit into this text. You can get an overview of the output by accessing **Usage Note 35968** mentioned earlier.

With all of these great sources of information about SAS products easily available, you can always have the most current information about what your organization has licensed.

Master the SAS Products Your Organization Has Licensed

Once you know what SAS products you have installed, you should ask yourself whether you know how to use them. If you do not know how to use them, learning them is a great way to continue to develop your SAS skills and achieve your goal of becoming a top SAS programmer. There are several ways that you can get familiar with SAS products you are not acquainted with.

SAS Documentation – The SAS Support website has numerous web pages devoted to SAS documentation. You can find information about individual SAS products (for example, SAS/GRAPH) and about particular procedures (for example, PROC GCHART) in the online documents. Online documents are usually available in both HTML and PDF format and have facilities for printing hard copies of sections. More information about SAS online documentation can be found in Chapter 5, "SAS Documentation."

SAS Press Books – SAS Press books are written by top SAS programmers in the user community and from SAS staff. Many of the books focus on a particular SAS product or feature and contain lots of code examples. The SAS code examples in these books can be found in the SAS Press pages at http://support.sas.com/publishing/authors/index.html. Users select an author, the book, and then the Sample Code and Data link. More information about SAS Press books can be found in Chapter 5, "SAS Documentation."

SAS Sample Library – Each SAS product comes with sample programs stored in a sample library. These programs are well-documented and provide examples of executing various procedures and facilities for the given SAS product. Sample libraries sometimes also contain SAS data sets that are used in the sample programs. You can find the sample libraries in the subdirectory for the product it is associated with. For example, on a Windows workstation, the sample library for SAS/GRAPH is located in this directory:

```
C:\Program Files\SAS93\x86\SASFoundation\9.3\graph\sample - 32-bit SAS
C:\Program Files\SAS93\SASFoundation\9.3\graph\sample - 64-bit SAS
```

Browse through the sample library of SAS products you are not familiar with. Open the programs, read the documentation, execute the programs, and look at the output. Do you understand how they work? If not, research the statements, procedures, options, and so on, in the related SAS documentation. Find out how they work and think about how you can use them in your assignments.

SAS Conference Papers – Every year dozens of SAS Users group conferences are held throughout the United States and in many other countries. Most conferences publish the technical papers that are presented, usually on the user group's website. Many of the technical papers provide overviews of SAS products and tutorials on how to use them. They are a good source of information about the SAS products and features you are not familiar with. More information about SAS conference proceedings can be found in Chapter 7, "SAS Users Groups."

SAS Classes – The SAS Education Division offers a wide variety of SAS classes that can help you learn about SAS software. They present e-learning web classes, prerecorded lectures by experienced SAS instructors, classes held in SAS training centers, and on-site training by SAS instructors. All of these options provide top-notch training you can consider taking advantage of to increase your knowledge of the SAS products you either do not know, or feel "iffy" about. More information about SAS classes can be found in Chapter 8, "SAS Training and Certification."

Once you know the SAS products and the releases of SAS that your organization has installed, and once you have some idea of how they work, you need to be thinking about *how* you can use them. How can you leverage these SAS modules to better input, store, process, and analyze data? How can you use them to better present the results of analysis? Think about your data and about your clients' needs. Can you use SAS/GRAPH to create eye-catching graphs that characterize result sets? Can you use SAS/CONNECT to access data from your Windows desktop that is stored on a UNIX server? Will you be able to use SAS/ACCESS Interface to ODBC to process tables in a Microsoft SQL Server database? Be creative. Take time to think about how you can best use all of the SAS tools available to you in ways they have not previously been used before in your organization.

Know Your Operating Environments

SAS does not operate in a vacuum. It runs interactively or in batch on a computer's operating system. Many facets of the operating system affect how SAS functions. Factors such as the version of the operating system, whether it is a workstation or a server, how files are allocated, how batch programs are submitted, and what else is running on the computer can have an effect on your SAS programs. Consequently, you need to take the time to learn some things about your operating environments in order to better facilitate the execution of your SAS programs on your organization's computer platforms.

Know the Version of Your Operating Systems

What version of Windows workstation software do you use to run your SAS programs? That is likely an easy question. Okay, so what version of Windows server, or UNIX, or Linux, or z/OS are you running SAS on? The answers to those questions might be less obvious. But kudos to you if *do* you know the answers off the top of your head!

Because of SAS' multi-vendor architecture, it can run on over two dozen UNIX, Linux, Windows, and z/OS operating systems. It is important for you to know the version of the operating systems you run SAS on so that you can understand the capabilities of those computing environments. Also, this is very pertinent information to have on hand when discussing issues with SAS Technical Support.

There are several ways you can determine the version of your operating systems:

Windows Command– Type the word *winver* in the Start window on Windows workstations and Windows servers. This results in a pop-up window specifying the current version of Windows, including the Build and the Service Pack. For example: *Windows 7(Build 7601: Service Pack 1)*

SAS Automatic Macro Variables – Within SAS, there are two automatic macro variables you can use to determine the version of your operating system:

- **SYSSCP** – This macro variable stores an abbreviation of the name of your operating system (for example, *HP_64* for HP-UX PA-RISC or H64 UNIX, *LIN X64* for LINUX on X64 (X86-64), *WIN* for Windows XP Pro, or *OS* for z/OS).
- **SYSSCPL** – This macro variable contains a longer abbreviation of the name of your operating system (for example, *HP_UX* for HP-UX PA-RISC or H64 UNIX, *Linux* for LINUX on X64 (X86-64), *XP_PRO* for Windows XP Pro, or z/OS for z/OS).

You can surface the value of both macro variables by using simple %PUT statements like this:

```
%PUT &SYSSCP;
%PUT &SYSSCPL;
```

Executing these statements in a Windows 7 32-bit environment produces this in the SAS log:

```
1     %PUT &SYSSCP;
WIN
2     %PUT &SYSSCPL;
W32_VSPRO
```

SAS programmers who write code for execution in multiple operating environments often use one of these macro variables to determine which operating system SAS is executing on. For example, this code determines whether the program is being run on Windows or UNIX, and allocates the data library Templib according to the appropriate file directory:

```
%Macro Allocate;

%if &SYSSCPL = W32_VSPRO %then %do;
      libname templib "c:\temp";
%end;
```

```
%else %if &SYSSCPL = HP-UX %then %do;
      libname templib "home\raithel\temp";
%end;
%Mend allocate;

%Allocate;
```

Talk to Systems Staff – This might be the old-fashioned way, but it works. Every organization has systems professionals who install, configure, and maintain its computers. They are a great source of knowledge about an organization's computers and operating systems. They also know about future upgrades to the hardware and software, when backups are taken, which time periods have the heaviest usage, and when planned outages are scheduled. So, they are a great source of information about the versions of your various operating systems as well as other pertinent software and hardware information. It is a good idea to be on a first-name basis with your organization's talented and knowledgeable systems staff!

Another reason for knowing the current version of your operating system is related to which specific versions of SAS it supports. All operating systems have a shelf life. They are released and then supported by the vendor for a number of years before a new version supersedes them. After a while, the vendor no longer releases hot fixes and maintenance releases for the old version, does not provide technical support for operating system issues, and it becomes harder to find and install the obsolete operating systems on new computers.

SAS software keeps pace with the evolution of the operating systems it is run on. Thus, periodically, SAS announces that it will not support obsolete operating systems with a new release of SAS. (For example, SAS discontinued support for the VAX operating system for all software releases after SAS 8.2). So, in order for you to migrate from your current version of SAS to the next, you need to know what version of the operating system you have. You must make sure SAS is supported on *that* particular operating system.

Fortunately, it is easy to determine which operating systems support SAS. The SAS Support website (http://support.sas.com) provides a comprehensive list of supported operating systems. You can find the list by selecting **Knowledge Base → System Requirements → Supported Operating Systems**.

Here is a screen shot of the Supported Operating Systems main page as of this writing:

Figure 3.4: Example of the Supported Operating Systems Page

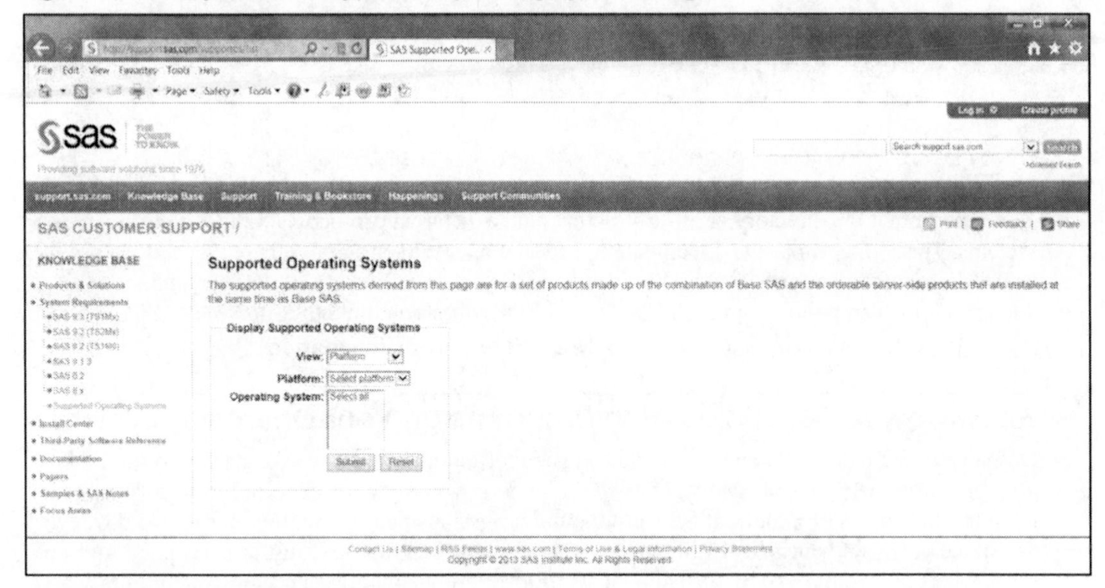

Know How to Allocate Files on Your Operating System

The Windows, UNIX, Linux, and z/OS operating systems have different file systems and consequently different syntax for how files and SAS libraries are specified. On Windows, UNIX, and Linux platforms, SAS libraries are simply directories that contain SAS data sets, catalogs, and other SAS objects. On z/OS, a SAS library is a file created by SAS with a data set organization of Physical Sequential (PS) and a record format of Fixed Standard (FS) that contains one or more SAS entities such as data sets and catalogs. It is obviously important for you to know the syntax of the operating system you are running your SAS programs on in order to correctly specify a particular SAS library or file.

In SAS, SAS libraries are usually allocated with a LIBNAME statement, and files other than SAS are normally allocated with a FILENAME statement. Here are examples of allocating SAS libraries in the Windows, UNIX, Linux, and z/OS environments, respectively:

```
libname winlib "c:\mikedir\prodlib";

libname unixlib "/home/mikedir/prodlib";

libname linuxlib "/disk1/mikedir/prodlib";

libname ZOSLIB  "PRODLIB.MIKEDIR.SASLIB" DISP=SHR;
```

These are examples of allocating files in the same four operating systems:

```
filename winfile "c:\mikedir\testfile.txt";

filename unixfile "/home/mikedir/testfile.txt";

filename linuxfile "/disk1/mikedir/testfile.txt";

filename ZOSFILE  "PRODLIB.MIKEDIR.FLATFILE" DISP=OLD;
```

You can see that the directory and file structures have different protocols on the various operating systems. The *SAS Companion* documentation for the operating system you are working on provides an excellent reference for how SAS libraries and files are specified. The *SAS Companion* documentation can be found in both HTML and PDF format on the SAS Support website (http://support.sas.com) by selecting **Knowledge Base → Documentation**.

Know How to Schedule Batch Programs on Your Operating System

Despite the fact that you are a hard worker, it is doubtful your manager expects you to run SAS programs from SAS Display Manager at 1:00 a.m. Nor would you be expected to stop your other work to run a series of important SAS production programs one after another during the day, every day of the week, and every week of the year. This is what batch processing was designed for. You run SAS programs in *batch* by providing a set of instructions to your operating system as to where to find the program, when to run it, where to store the output files, and sometimes which other programs should be executed next. Different operating systems have different ways of scheduling SAS programs to execute in batch. The following sections describe batch scheduling in the most prevalent operating systems. Learn to use the batch schedulers on your organization's operating systems so that you can be soundly asleep at 1:00 a.m. when your SAS programs are running.

Windows

On Windows, you create a .bat file and use the Windows Task Scheduler. The .bat file is simply a text file where you put the full path of the SAS executable, the path to the SAS config file, the name of the program to be executed, destinations for the log and list files, and other pertinent SAS options that can be specified at SAS invocation time. For example, the .bat file C:\Big Project\Production Programs**Job1.bat** could contain the following instructions:

```
"C:\Program Files\SAS93\x86\SASFoundation\9.3\sas.exe"
-icon -noterminal -nosplash
-noxwait -noxsync
-CONFIG "C:\Program
Files\SAS93\x86\SASFoundation\9.3\nls\en\SASV9.CFG"
-SYSIN "Q:\programs\Driver_Program.sas"
-LOG "Q:\programs\Driver_Program.log"
```

The instructions specify that when the .bat file is submitted to the operating system, SAS is to execute the Driver_Program.sas program in the **Q:\programs** directory and to store the log in that same directory. The .bat file specifies a number of SAS options as well as the location of the SAS config file.

You access the Task Scheduler on Windows by selecting **Start→All Programs→Accessories→System Tools→Task Scheduler.**

Here is an example of the Task Scheduler under Windows 7 Professional Edition:

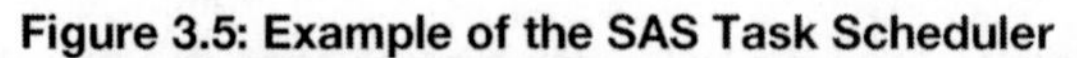

Figure 3.5: Example of the SAS Task Scheduler

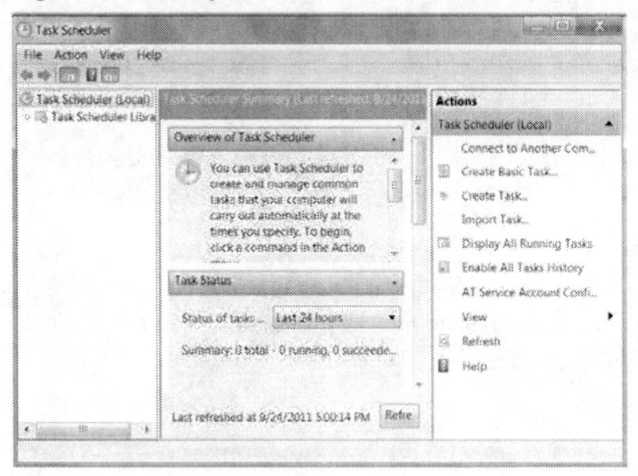

It is beyond the scope of this text to provide detailed information about how to use the Task Scheduler. But it is easy enough to learn how to use Task Scheduler, and the Help menus provide good reference material.

UNIX

In UNIX environments, SAS can be run in batch by creating a script file and then using the crontab to schedule when the script file is executed. The script file is a text file that contains instructions for UNIX to execute SAS and identifies where the program is located, what SAS initial options should be run, and where the log and list should be stored. For example, the script `/home/unixacct/programs/MONTHLY_UNIX_ACCOUNTING_DRIVER_SCRIPT` contains the following instructions:

```
################################################
# Script: MONTHLY_UNIX_ACCOUNTING_DRIVER_SCRIPT #
#                                              #
# Author: Michael A. Raithel                   #
#                                              #
# Created: 04/27/2020                          #
#                                              #
# Purpose: This script processes the monthly   #
# Unix Accounting files into a SAS data set    #
# and creates monthly reports.                 #
################################################
/home/sas93/sas /home/unixacct/programs/monthly_unix_accounting.sas \
-noterminal -log /home/unixacct/programs/monthly_unix_accounting.log \
-print /home/unixacct/programs/monthly_unix_accounting.lst
```

When this script is submitted, it executes the monthly_unix_accounting.sas program and saves the log and list files to the same directory where the program is located. Refer to a UNIX reference book to read more about UNIX script files.

The crontab is a native UNIX utility that enables you to schedule the execution of UNIX commands and scripts in the background by the crontab daemon. The crontab has five fields you can use to specify when something is executed:

- Minute: 0–59
- Hour: 1–24
- Day of Month:1–31
- Month: 1–12
- Day of Week:0–6, where 0 = Sunday

The date/time fields are followed by the particular script or command to be executed. For example, this crontab entry specifies that the indicated script is to be run at 7:30 p. m. on the first, fifteenth, and thirtieth day of January, April, July, and December.

```
30  19  1,15,30  1,4,7,12  *  MONTHLY_UNIX_ACCOUNTING_DRIVER_SCRIPT
```

You can edit, display, and remove your crontab file via the following crontab commands, respectively:

```
crontab -e

crontab -l

crontab -r
```

There is a lot more to know about crontab files, but more detailed explanations are beyond the scope of this text. However, you can find those details in the man_pages on your UNIX system. Simply enter the following at the command line on your UNIX server: man crontab.

z/OS

In z/OS environments, batch programs (referred to as "batch jobs") are submitted by using Job Control Language (JCL). The JCL tells z/OS what system resources to allocate and which SAS programs to execute. A batch job can consist of one or more "steps." Each step executes a separate program. When an individual step of a batch job executes, the SAS program specified in the step processes data. System utilities and programs written in other programming languages might also be executed in separate steps of a batch job.

Here is an example of JCL:

```
//DAILYDBL  JOB  (LEDZEP,7422),'LEDZEP DAILYDBL',
//      MSGCLASS=X,MSGLEVEL=(1,1),
//      CLASS=5,NOTIFY=BBZE0LK
//*****************************************************************
//* LOAD SONG TRANSACTION DATA INTO THE MUSIC DATA WAREHOUSE.     *
//*****************************************************************
//STEP01 EXEC SAS
//DATA      DD DSN=LEDZEP.PROD.DATAENTY.DATA.SASLIB,DISP=SHR
//TRANFILE DD  DSN=LEDZEP.PROD.CDSALES.TRANFILE,DISP=SHR
//SYSIN     DD DSN=LEDZEP.PROD.SOURCE(SONGEXTR),DISP=SHR
//*
//*****************************************************************
//* DELETE DATA SET  WITH IEFBR14                                 *
//*****************************************************************
//STEP02   EXEC PGM=IEFBR14
//SYSPRINT DD SYSOUT=*
//TRANFILE DD  DSN=LEDZEP.PROD.CDSALES.TRANFILE,
//          DISP=(MOD,DELETE,DELETE),
//          UNIT=SYSDA,SPACE=(TRK,(1,0))
//*
```

The JCL in this batch job executes two steps. The first step executes a SAS program stored in the SONGEXTR member of the LEDZEP.PROD.SOURCE partitioned data sets (PDS). The second step executes an IBM utility that deletes the LEDZEP.PROD.CDSALES.TRANFILE data set.

In most organizations, you can submit batch jobs from your mainframe computer sessions to execute immediately. However, if you want to execute a batch job on a cyclic basis (hourly, daily, weekly, monthly, and so on), then you need to schedule it via a job scheduler.

Job schedulers are sophisticated software programs running 24/7 that automatically submit and monitor batch jobs. They execute batch jobs at the specific times they are scheduled for, route the output to specific queues, submit other batch jobs that are dependent on the execution of the first ones, and alert programmers to failed batch jobs. Organizations usually have production management staff who use back-end job scheduler software to set up when a batch job executes, its dependencies, where the print output goes, and who is contacted if the job fails. There is usually a nice bureaucracy built up for filling out forms and having them signed to authorize your batch job to be officially entered into the batch job scheduler for execution.

Talk to the systems staff in your organization and find out what the procedures are for scheduling a job to run on z/OS using your organization's batch scheduler. Learn what paperwork (electronic or paper) you need to fill out, what authorizations you need to get, whether your SAS programs and JCL need to be stored in official "production libraries," what job classes are relevant to running your batch jobs, what account information you need on your JCL's job statement, and any other details you must know in order to have your jobs scheduled for execution. Then, take a load off of your personal schedule by moving important, cyclic batch jobs to the z/OS batch scheduler for execution.

Know the Nuances of Your Operating System That Affect Your SAS Work

Despite your best efforts to be productive, there are still some dynamics of your operating system that can adversely affect your SAS work. Most of these influences are a result of working in a collaborative data processing community where computer resources are shared among a small, medium, or large number of colleagues. When many of your colleagues are using the shared computer resources, everybody's programs run slower. Competition for computer memory, input/output resources, temporary work disk space, and data transfer bandwidth are some of the many factors that can slow down the execution of your SAS programs.

If you run SAS programs in a Windows server, UNIX, Linux, or z/OS operating environment, then you are probably very aware of the fact that "the more" is not "the merrier" when it comes to using those operating systems. If you are doing most of your work on a PC, then it might not be readily apparent that more users can affect the elapsed time of your SAS programs. However, it is likely that some or all of your production data resides on network directories accessed via a Local Area Network (LAN). When other staff members are transferring data across the network to and from their own PCs at the same time you are, the sheer volume can slow down everybody's overall processing time. Competition for corporate computing resources can be a problem when you have a lot of programs to run and a rapidly approaching deadline. Consequently, it is a good idea to identify some of the dynamics of your operating system that could adversely affect the timely completion of your SAS programs.

Here are some of the factors you should consider:

Many Concurrently Executing Tasks – When many programs are executing at the same time on a shared resource such as a UNIX, Linux, or z/OS computer, they vie for access to computer resources such as memory, I/O, and CPU. Operating systems have rules for swapping concurrently executing programs in and out of memory to best maximize systems resources. But no matter how clever the operating system is, an overabundance of concurrently executing programs results in all programs taking longer to execute.

System Backups and Database Backups – Most organizations take backups of system disk storage devices and databases at night or early in the morning. A usual scheme is to take a full disk or database backup once a week and then make an "incremental" backup (only backup what has changed) during the week. When backups are running, a lot of data is being transferred over the network between the disks and the backup medium. So, if you run your SAS programs while backups are running, you will likely see elongated execution times as they compete for network transfer bandwidth with the backup tasks.

Slow Network Performance – Sometimes the network just seems slow. This could be due to network problems, but it is most often due to many people using it at the same time. Like you, they are running interactive and batch programs, transferring files, opening spreadsheets, sending e-mails, and doing a host of other productive tasks. All of this concurrent activity puts a strain on your network's bandwidth resources and leads to slower performance for tasks that normally wouldn't take very long.

Slow Performance on Certain Servers – In a multi-server environment, some servers can experience slower overall performance than others. This could be due to many factors. Issues such as the server having slower processors, less computer memory, slower disk transfer speeds, and smaller disk work areas can affect the performance of a server. Some frugal organizations hold onto older computer systems to wring every penny out of their investment, so some computers and their operating systems might be old. And some servers might simply be very busy with an overabundance of users. Whatever the case, if you work in a multi-server environment, it is important for you to identify such servers and determine whether there are others you can use instead to get your work done.

Actively consider how the factors described above might affect your ability to run SAS programs in your operating environments. Determine the best and worst times to execute your programs:

- **Business hours, 9:00 a.m. to 5:00 p.m.** – These are usually the "rush hours" during which a lot of people are doing work that vies for your organization's computer resources. Nobody expects you to only work at night or on the weekends to avoid this phenomenon, but be aware that it exists. In many organizations, the time from 12:00 p.m. to 1:00 p.m., when most people go to lunch, is a good time during the business day to run programs.
- **Evenings, early mornings, and weekends** – These are usually the best times to run your very resource-intensive SAS programs, the ones that need a lot of computer memory or CPU time, or that process large files. Times outside of the core business hours historically see lower demands upon computing resources. So take advantage of these times, especially with your more resource-intensive SAS programs.
- **Backups** – Avoid the times when backups are being taken. Because backup jobs consume a lot of network bandwidth, they will retard the execution of your SAS programs, which elongates the execution time. Contact your systems department and your database administrators. Ask them when they take network and database backups, respectively. Make note of those times and try to schedule your off-hours SAS work around them.

- **Big production programs** – Most organizations run big production programs, such as payroll or corporate accounting, on a cyclic basis: daily, weekly, monthly, quarterly, etc. Such big production programs can consume large amounts of computer resources and might also put a strain on network bandwidth. Learn the schedule for these programs if they run on servers that you run your own SAS programs on. To the extent possible and practical, avoid running your own SAS programs during those periods.

Knowing the nuances of your operating system that affect your SAS work can make you more productive. It can help you become more strategic in determining when to run your SAS programs and in choosing what servers you run them on. Most importantly, it can help you to avoid a situation where you miss a deadline for an important deliverable because your SAS program was delayed due to slow computer resources.

Chapter 4: What You Should Know at Work

Introduction

You will probably not be very surprised to read that you need more than just a mastery of SAS to become a top SAS programmer. Mastering SAS is a good first step, but knowing everything about your programming tools will get you only so far. Consider a carpenter with a toolbox full of the latest tools who knows exactly how each tool should be used to build a home. That carpenter also needs to know about the building materials the tools will be used on. The carpenter needs knowledge of two-by-fours, plywood, subflooring, rafters, sheetrock, molding, joists, plaster, cinder blocks, nails, screws, window frames, and so on. The carpenter also needs to understand exactly what the client wants to have built. So the carpenter must know how to read blueprints, understand local building ordinances, know how to obtain building permits, and understand how to interpret his client's building specifications. A simple mastery of the tools, alone, is not sufficient.

These concepts are also true in the world of SAS programming. You need to have an in-depth knowledge of the data you are going to be using in your SAS programs. You must understand the data stored in your organization's Excel spreadsheets, Microsoft Access databases, comma-separated value (CSV) files, fixed-length or variable-length flat files, tab-delimited files, XML files, relational databases, and HTML files to name a few. The better you know the data, the better your ability to analyze it and present cogent results to your clients. In addition, you also need to have a very firm grasp of what, exactly, your clients want. Whether they are internal clients (such as your manager or analysts in another group) or external clients (such as a federal agency your organization does business with), understanding their specifications and needs is vital to helping you and your organization be successful.

This chapter discusses two basic areas that you should be knowledgeable about at work. The first is that you should know your data. The second is that you must know your client's requirements. Mastering these two areas will help you continue on your way to becoming a top SAS programmer.

Know Your Data

Data are the lifeblood of an organization and comes in an amazing variety of topical areas. There are personnel data, corporate accounting data, payroll data, corporate assets data, EOE compliance data, capital expenditures data, security compliance data, survey response data, and the data collected on behalf of doing business with your organization's clients. It is your job as a programmer to access, process, analyze, and present your organization's data. To do that, you must have an understanding of the characteristics of the data you are working with. A comprehensive understanding of the data enhances your ability to analyze it and to wring out accurate, insightful, and actionable information.

This section provides several data-related topics that you should become familiar with in your organization. The topics range from knowing where the data are located to understanding the nuances of the data. All of the topics are important to your success in wielding SAS to create the analysis results your users need.

Know Where the Data Are Located

An obvious first question when starting any assignment is, *Where are the data located*? You need to know which directories the data sets are stored in. If the data are in a database, then you need to know the name of the particular database and the specific tables within the database that you must access. For data on removable storage, you must know which CD or DVD the data were written to. When dealing with tapes, you need to know which tape volume or volumes the data are spread across. Whatever the storage medium, either the person who assigned the task or a corporate data administrator should provide you with clear information about where the data can be found.

Another consideration is whether the data are immediately accessible. Older data sets might be archived to tape and stored off-site. If that is the case for files you need, you can expect a certain amount of lag time as your request for them is processed, the files are retrieved, and they are written to a network directory you have access to. If the data are stored on a mainframe or UNIX server, then you need to determine whether you have a valid account to access the server. After you get the access issue sorted out, you must figure out which directories or data sets house the data you need.

Once you know where to find the data, you must establish whether you have the correct level of permissions to access them. Data security has become an increasingly hot topic, resulting in most organizations tightening the security controls protecting access to their data. It is now common practice to lock down access to data by default, and to only grant access on a need to know or a need to use basis. Data on networks are usually protected by network software which grants access rights to individuals based on permissions given by security administrators. Access to data in databases is often governed by database administrators who grant Read, Write, and Update access. Some organizations have a unit responsible for processing requests for data access, while others rely on the data owners in the individual areas of the organization.

So you must make sure the security apparatus in your organization grants you the proper authority to access the data you need for your assignments. This might be as simple as sending an e-mail request to a project manager or a Help Desk. Or, it might be as complicated as filling out online forms or signing hard-copy confidentiality agreements. Whatever the case, it is a good idea for you to do this up front when you first get wind of an assignment. Doing so enables you to navigate through the bureaucratic data access request process in a timely fashion so that your programming assignments are not unduly delayed.

Know the Format of the Data

Another consideration is the format in which your data are stored. Are the data stored in SAS data sets, Excel files, a Microsoft Access database, or an XML file? Are the data housed in a relational database such as SQL Server, Oracle, or Sybase? The format the data are stored in affects the SAS tools you can use and the amount of effort required to access it.

If your data are stored in SAS data sets (files with a .sas7bdat extension), then you are in luck because SAS understands SAS data sets best of all! SAS knows whether variables are character or numeric, the length of each variable, formats and informats, variable labels, whether the data sets are sorted, the size of the data sets, when they were created, when they were changed, and more. You can bring the full power and might of SAS to bear when processing SAS data sets with few reservations. However, there are a couple of issues to be aware of.

- Data sets might be in a foreign data representation[1] if they were simply copied or transferred using FTP from another operating system. In that case, SAS uses Cross Environment Data Access (CEDA) engines to process them, resulting in slightly slower processing times and restrictions on some operations such as updating data sets in place. You can fix that by simply using PROC COPY to make copies of the data sets in another directory. As SAS copies the data sets, it "flips the bits" so that the resultant copies are in the same format as the operating system they are stored on. You can determine the data representation of SAS data sets by running PROC CONTENTS and selecting the `Host Created` field in the Engine/Host Dependent Information section.
- Another issue arises when the SAS data sets have user-created formats. When they do, you need to assign the catalogs those formats are stored in to the list of format catalogs SAS searches when executing your programs. This is usually done via the FMTSEARCH, INSERT, or APPEND options. Failure to do so results in SAS surfacing an error message every time you attempt to open the data sets. That message states that the format cannot be

found and the data set will not be opened. You will not be able to open the data set until you use one of the aforementioned options, or you use the NOFMTERR option to open the data set without the format.

If your data are stored in SAS transport files, then you are still fortunate because transport files can be transformed back into SAS data sets with very little effort. Transport files are linear SAS files used to transport SAS data sets between operating systems or between versions of SAS. There are two types of transport files: those created with PROC COPY and those created with PROC CPORT. You can use PROC COPY to re-create SAS data sets from transport files created by PROC COPY. Use PROC CIMPORT to re-create SAS data sets from transport files created by PROC CIMPORT. (If you are not sure which type of SAS transport file you have, read SAS Usage Note 22656 or see the edition of the SAS publication *Moving and Accessing SAS Files* for the version of SAS you have installed).

SAS can process data residing in Microsoft Access databases or Excel spreadsheets via SAS/ACCESS Interface to PC Files. You can verify whether you have it licensed by simply running PROC SETINIT and checking the list of products written to the log. If you do have SAS/ACCESS Interface to PC Files licensed, then you are in luck and can easily read from and write to Microsoft Access and Excel files. You can do so via the SAS Display Manager's Import and Export wizards, or with specifications in the LIBNAME statement, or with the IMPORT and EXPORT procedures. If you do not have it licensed and will be doing a lot of work with Access and Excel files, look into obtaining this very helpful SAS software. It will make your programming life a lot easier!

The Extensible Markup Language (XML) has become a popular standard for sharing data across the Internet and between operating systems. So it is very likely that sooner or later you will be called upon to either read from or write to an XML file. Fortunately, XML files can be processed with the SAS XML LIBNAME engine. This Base SAS product enables you to import an XML document into a SAS data set, and to export a SAS data set into an XML document. For information about how you can access and create XML files with SAS, see the latest edition of *SAS XML LIBNAME Engine User's Guide.*

Data stored in databases can also be accessed by a variety of SAS products. Here are some of them:

- SAS/ACCESS products that facilitate access to relational databases include:
    - SAS/ACCESS Interface to DB2
    - SAS/ACCESS Interface to Informix
    - SAS/ACCESS Interface to Microsoft SQL
    - SAS/ACCESS Interface to MySQL
    - SAS/ACCESS Interface to ODBC
    - SAS/ACCESS Interface to OLE DB
    - SAS/ACCESS Interface to Oracle
    - SAS/ACCESS Interface to Sybase

- SAS/ACCESS products that permitting access to non-relational databases include:
  - SAS/ACCESS Interface to CA IDMS
  - SAS/ACCESS Interface to ADABAS
  - SAS/ACCESS Interface to CA-Datacom/DB
  - SAS/ACCESS Interface to IMS
  - SAS/ACCESS Interface to SYSTEM 2000

These SAS/ACCESS products enable you to read, update, and create database tables in your SAS programs. You can supply "connection string" information about a LIBNAME statement and then process the tables in the databases much the same way as you would process data sets in a SAS library. Some of these SAS/ACCESS products facilitate in-database processing of SAS procedures such as PROC SORT and PROC MEANS. So, the database engines do the work inside the database and then return the result sets to SAS. In-database processing can speed up your SAS programs and reduce their network bandwidth when only the smaller result sets are returned from the database to your programs. Consequently, it is well worth looking into the feasibility of using one of these SAS/ACCESS products with the particular DBMSs you are accessing.

Determine whether you have the requisite SAS products that enable you to process data stored in your organization's relational databases. If you do not, and will be doing a lot of work with relational databases, you should seriously consider purchasing the related SAS/ACCESS product. SAS/ACCESS products give you the autonomy to explore and process tables in relational databases at will. Otherwise, you will have to rely on a database administrator or database programmer to provide extract files for you to process.

Know the Layout of the Data

You cannot process your data until you understand the data's layout. You need to know what constitutes a record. Is it a row in a database table, or rows joined between several database tables based on a common key? Is a record a single row in a flat file or multiple rows identified by a key variable? Is a "record" a row in an Excel file or an observation in a SAS data set? You need to recognize what, exactly, constitutes a "record" before you can process a record.

Once you identify what constitutes your records, you must understand what constitutes a variable. If you are processing SAS data sets, relational databases, Excel or Access files, then this is straight-forward. These software products have clear rules for what makes a variable or a column. The rules are often reinforced with metadata that you can inspect to understand the characteristics of the data. It is not usually so obvious with flat files, unless they are delimited with column separators such as tabs, commas, or pipes. Flat files require documentation that specifies where variables begin and end, and whether the file is fixed length or variable length.

After you understand what constitutes your variables, you need to identify their characteristics. What are the variable names? Are the variables character or numeric? What are their lengths, formats, and labels? What constitutes a missing value? If numeric variables are not integers, what is the decimal precision? The metadata for the individual variables is also important information for you to have in order to truly understand the data.

These issues are some of the things you must understand about your data. Obtain documentation such as data dictionaries, PROC CONTENTS, code books, and so on, for all of the data you are going to be working with. Read through the documentation and thoroughly understand the layout of the data. Doing so enables you to be nimble when using SAS to wrangle information out of your organization's data.

Know the Nuances of the Data

Data without context is meaningless. To create meaningful analysis, you need to understand what your data represents. Your organization's data records could represent customer orders, survey responses, hospital visits, drug dosages, hours worked, or a host of other measures and events. So you need to learn what the records in your data sets or the rows in your relational databases represent.

You can get information about what the data represents from a number of sources. This information is often found in system documentation and codebooks. Consequently, your best bet for obtaining this material is to talk to system administrators or the subject-matter experts who set up the systems that house the data. They are bound to have formal documents specifying the details for how data is collected and stored, and what it represents. Those documents and discussions with the aforementioned staff are your keys to understanding the data.

There are also nuances to the variables that exist in your data sets and database tables. For example, there are different types of variables:

- **Key variables** – These are one or more variables that uniquely define a record in a data set. Sometimes a key variable can be an identifier such as a patient number and there are multiple records with that value in the data set. Other times, it is permissible to have only one record with the unique combination of key variable values.
- **Categorical variables** – These are qualitative variables that are used to categorize your data. There are two types:
    - **Nominal variables** – These variables represent categories with no particular order (for example, Male and Female).
    - **Ordinal variables** – These variables represent categories that can be ordered, such as income categories or the degree to which one agrees with a question (for example, 1=somewhat agree, 2=agree, 3=strongly agree, and so on).
- **Discrete variables** – These values are specific and represent a count, such as number of doctor visits.
- **Continuous variables** – These values represent a measure, such as weight or height.

- **Interval variables** – These variables represent intervals in time during which something happened.
- **Composite variables** – These are variables derived from other variables within a data set or from variables that exist in multiple data sets.

The allowable values of variables are also important. You should know if and when missing values are acceptable for the variables in your data. For categorical variables, understand what the values (for example, 1, 2, 3, and so on) represent. Know when groupings of variables within your records represent arrays. Understand whether certain values denote special events or occurrences. For example, "-9" might represent that the respondent to a questionnaire refused to answer. Be aware that there might be flag variables that signal whether other sections of variables are present or missing.

Another nuance of your data is its periodicity. If you are processing data in a database that is continuously updated, then today's analysis might produce different results from last week's. Many organizations have analysis performed on a data warehouse or on shadow databases that are periodic extracts of the operational databases. If this is what is happening in your organization, learn when the data are refreshed. If you are going to be performing a long-term analysis on the data, perhaps you need to create your own extract as a SAS data set. If not, then you need to at least understand the cutoff date for the data collection that you are processing.

Know Your Clients' Requirements

In addition to your supervisor and your colleagues, your clients are the most important people in your work life. They are your true customers and more than likely the originators of the core IT tasks you perform on a daily basis. They are often your best advocates when you are meeting their needs and can be your sharpest critics when you are not. Your clients might be internal users such as business analysts in another part of your organization. Or, they might be staff in an external business that has contracted with your organization for data processing and analysis. Keeping your clients happy is imperative to the success of your organization and to your own chances for success and career growth. So, a solid understanding of your clients' requirements is of paramount importance!

The following sections discuss some of the issues you should consider in order to gain a better understanding of your clients' requirements.

Know the Specifics of the Assignment

Many organizations have a specific regimen for how assignments are made available to the programming staff. Sometimes an assignment request form is filled out, sometimes an analysis memo is presented, sometimes detailed specifications are drafted and submitted, and sometimes it is simply encapsulated in an e-mail. You are probably familiar with how it is done in your organization and how your own clients request assignments. No matter the vehicle for the assignment, here is some of the most important information for you to know:

- when the assignment is due
- how "urgent" this task is in relation to other tasks you are working on
- the data sets or databases that should be used for the assignment
- the selection criteria for the data
- the formulas for any calculations that are to be computed

In addition to these elementary issues, understand how your clients want the results to be delivered. There are considerations for various types of results:

- **Data sets** – If the deliverable is data sets, you must know whether they should be SAS, Excel, CSV files, ASCII text files, or even new database tables. Make sure you know what your clients are expecting for the order of variables in the data sets. Also find out if your clients have specifications for variable names, variable labels, variable types, and variable formats.
- **Reports** – You should understand the layout of the report yours client are expecting. Are they expecting basic reports or something flashy, such as can be created using advanced features of the Output Delivery System? Know if your clients need reports on a special letterhead, or if there are official titles or footnotes that should be on every report.
- **Graphs** – Know what your clients are expecting for the various axis, dimensions, reference lines, and labels of the graphs. Make sure you understand what is expected for title lines and footnotes. Also, know how many graphs are to be put on each page and the arrangement of graphs.
- **Documentation** – Find out whether your clients expect PROC CONTENTS or other types of data dictionaries. Sometimes SAS clients want the SAS logs delivered to them with their result sets.

Make it a point to get your clients' specifications in writing. It is best to start with written specifications on the basic assignment. The more detailed the specifications, the better. Keep notes from telephone discussions and meetings so that you can modify what your client is asking for. Then, send your understanding of clarifications and changes back to the client via e-mail so that everybody can see them in writing and verify the proposed changes. When possible, ask your client to ratify the specifications with any additional information that surfaces in discussions. Incorporating the changes into the specifications helps everyone stay on track with the latest thinking about the assignment. Overall, ask questions, take notes, and stay on top of the task as it evolves.

Understand Your Client's Needs

Take the time to make sure you understand what is really important to your client. Perhaps your client needs you to expedite all ad hoc reports because they are necessary to address government or regulatory oversight requests. Maybe your client is willing to wait a bit longer for an analysis to give you the proper time to perform extensive quality checks on the results. Perhaps your client prefers to have data delivered in SAS data sets or in Excel files. Maybe your client needs preliminary results immediately and the details later on. Perhaps the client likes to get weekly reports describing a project's status. Maybe your client needs all files to be zipped and password-protected before delivery. Perhaps *exact* deadlines are important to your client.

The better you understand your client's needs, the more you will be able to meet them. Good *customer service* on your part builds trust and a good working relationship with your client. It enables your client to be successful in the use of the information you are providing. Good customer service is essential to you and your organization maintaining a solid working relationship with your clients. So put the time and effort into learning what factors are the most important to your clients.

How Can You Be Proactive?

Many programmers engage in "stimulus-response" type programming. The client provides programming specifications, the programmer writes the programs that meet the specifications and returns the results to the client. There is nothing wrong with this basic scenario, as that is the normal relationship between programmers and their users. However, to become a top SAS programmer, you should go the extra step and anticipate your clients' needs. Try to think about what you can do for them between assignments or to enhance your deliverables:

- Can you provide additional analysis that your clients have not yet thought of which would add additional insights into the data?
- Can you create extract files that whip the data into shape for future analysis, so the data are ready for quick ad hoc reports?
- What about experimenting with ODS Statistical Graphics to enhance the next set of reports with graphs?
- Can you spice up your reports with better titles and footnotes and use ODS to create the reports in RTF?
- Is it possible to turn those tired old PROC PRINT reports into state-of-the-art reports using PROC REPORT?
- How about delivering the data in both a SAS data set and a CSV file for users who are not very savvy using SAS?
- Would it make sense to supply a CONTENTS procedure listing with your SAS data sets so that your clients don't have to create them?
- Can you restructure the reports you create to supply summary results in the first pages and details in subsequent pages?
- Would your users like to have the variables in a SAS data set ordered in a particular way, and the data set sorted by a specific set of variables?

You can probably think of dozens of ways to add value to what you are doing for your own clients. Don't just think about them; take action and do them. Surprise your clients by staying one step ahead of them in anticipating and meeting their needs. Show them that you are indeed the top SAS programmer in your organization.

---

1 SAS data sets are created using the data representation format native to the operating system they reside in. Such representations differ between operating systems. For example a Windows SAS data set has a different underlying data representation from a UNIX SAS data set. So, from the Windows perspective, a UNIX SAS data set has a foreign data representation.

Chapter 5: SAS Documentation

Introduction

Albert Einstein is credited with stating, "*Know where to find the information and how to use it—that's the secret of success*". And, I doubt that any SAS programmer would argue with one of the most famous winners of the Nobel Prize in Physics. Fortunately, SAS is very well documented in a number of convenient media, so it is relatively easy to find detailed information about how to install, configure, and use it. As a matter of fact, there is so much documentation and information available, that it can sometimes be a challenge to sift through all of it and determine the best places to look for specific information. This chapter provides an overview of the best sources of SAS documentation. You can find SAS documentation on the SAS support website, in the SAS Help menus in SAS Display Manager, in SAS technical references, and in SAS Press books.

Documentation that you find on the SAS website and in the SAS Help menus tends to be more formal in describing SAS features and usage. Such documents are written by SAS staff who are closely tied to the products they are writing about. This is also the case for the hard-copy SAS technical references, which you can order via the SAS website. Documentation found in SAS Press books tends to be a bit less formal, though just as informative. SAS Press books are written by SAS subject matter experts throughout the world and edited by SAS staff for technical accuracy.

All of the sources of SAS documentation are readily available to you. SAS online documentation and the SAS Help menus in SAS Display Manager are only a few mouse clicks away. You can order SAS technical references and SAS Press books on the web or at a SAS users group conference and receive them in the mail a few days later. So, as you read this chapter, start thinking about how you are going to create your own library of SAS documentation. Perhaps you will bookmark a number of SAS website pages or SAS users group web pages. Maybe you will save some of the SAS Display Manager help items in your Favorites tab. Perhaps you will order some key SAS Press books and SAS manuals. More than likely, it will be a mix of "all of the above", as you build both a soft-copy and hard-copy library of SAS documentation.

SAS Online Documentation

SAS online documentation is available to you 24/7, 365 days a year (366 days in leap years) on the SAS support website. And, it is absolutely free! On the documentation web pages, you can find everything from *Base SAS Procedures Guide*, to *SAS Statements: Reference*, to *SAS/STAT User's Guide*, to *Output Delivery System: User's Guide*. Once you find the publication that you want, you can peruse the contents as either HTML pages or as PDF pages.

SAS online documentation can be found at the following URL:

http://support.sas.com/documentation/

That page is your jumping-off point for finding SAS documentation. It offers many different ways you can go about searching for SAS facts, depending on your own inclination. Here is a screen shot of the Product Documentation web page as of this writing:

Figure 5.1: Example of the SAS Documentation Home Page

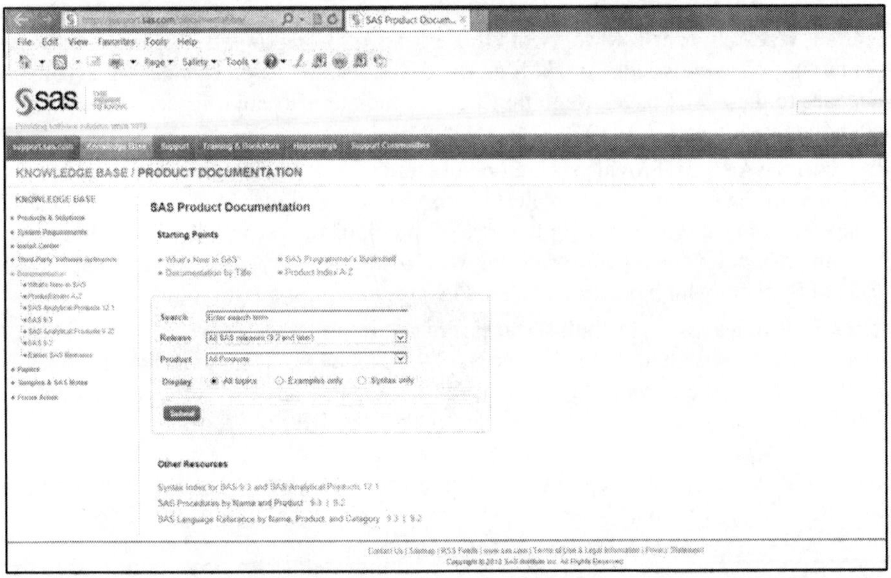

As you can see, there is a tree view on the left side and choices for you on the main page. We will discuss the tree view first and then the choices on the SAS Product Documentation main page.

The SAS Documentation Tree View

The tree view presents the following items you can click:

- **What's New in SAS** – This selection opens a window that enables you to look at the features introduced in the current release and in several previous releases of SAS. The What's New pages are *must-reading* when a new release of SAS is first issued. You should examine those pages and become familiar with the many new enhancements included in the new release. Doing so keeps you on the cutting edge of SAS and enhances your SAS programming technical capabilities.
- **Product Index A-Z** – This link brings you to exactly what you would expect, a list of SAS products in alphabetical order. Everything is in there! If you know the product you are looking for, simply find it in this list. One tip to remember is that you can disregard the word "SAS" when looking for a particular product and simply look for its specific name in the list. For example, "SAS/AF" is found in the "A" section, not in the "S" section. SAS/FSP is found in the "F" section, and so on. So don't look for "SAS"; just look for the product itself.
- **SAS 9.3** – This link surfaces a page that mirrors the basic SAS Documentation home page, except that it is slanted specifically toward the newest release of SAS. A more detailed description of the features of the main part of this page can be found in the next section of this chapter, "SAS Product Documentation Main Page."

- **SAS Analytical Products 9.22** – Documentation for SAS' premier analytical products such as SAS/ETS, SAS/IML, SAS/OR, and SAS/STAT can be found here. If you work with statistics or operations research, or if you work with those types of professionals, this is a great resource to go to for cutting-edge knowledge on SAS products and features. You can also refer your statistician colleagues to these web pages when they need to look up particular facets of these software applications.
- **SAS 9.2 and SAS 9.3**– Provide starting points for researching features of previous releases of SAS. Hopefully, you are using the latest release of SAS and these links are irrelevant. If not, then this is where you can go to look up features of previous releases until you can upgrade. These links also come in handy when you are working with SAS clients who are not working with the latest release of SAS but who have a particular SAS programming question for you.
- **Earlier SAS Releases** – This link surfaces a page presenting documentation for SAS 9.1.3 and earlier releases. As with the SAS 9.2 and SAS 9.3 link pages, it can be useful if you are working with clients who have *archaic* versions of SAS. It would be best to coax such clients onto the latest release, but of course that is their decision to make.

That wraps up the choices from the tree view. Make sure you open your web browser, navigate to the website, and become familiar with it.

SAS Product Documentation Main Page

The SAS Product Documentation main page offers four Starting Points, a search engine, and links to additional documentation resources. Each of these is described, in turn.

- **What's New in SAS** – This link takes you to the same web page as the first selection in the tree view. From there, you can see what is new in the latest release of SAS, or what was new in several previous releases. Make sure you are the first person in your organization to know what's new in the latest release of SAS. Then, make a point of disseminating that knowledge to your colleagues so that they learn it from you and put it to good use.
- **Documentation by Title** – You can use this link to find a list of SAS documentation alphabetized by title. It is a convenient way to find SAS technical references when you know what you are looking for. This page can be useful for just browsing through all of the SAS documentation titles to see what might be an interesting read. Look through the technical references for products your organization licenses and see if you are familiar with all of the procedures, statements, and options. Take a look at some of the technical references for products that your organization does not license and determine whether they might be beneficial.
- **SAS Programmer's Bookshelf** – This selection surfaces a very intelligent list of the resources SAS programmers are most likely to reference. It is a subset of the overall available set of SAS technical references. However, you are likely to find that most of the things you look up are in one of the documents linked on this page. It is definitely worthwhile to bookmark the SAS Programmer's Bookshelf page or to have a shortcut to it on your desktop.
- **Product Index A-Z**– This link takes you to the same page as the Product Index A-Z tree view link.
- **Search Engine** – The central part of the Product Documentation web page contains a powerful search engine that enables you to search SAS documentation for specific terms. You enter the term

in the Search window, select the release of SAS from the Release drop-down menu, and then select one of the following:

- **All topics** – The default is to surface every document source the particular search term is used in.
- **Examples only** – Only yields examples of the particular SAS construct you are searching for.
- **Syntax only** – Only produces the syntax for the SAS element you are searching for.

Performing a search produces two tabs on the search results pages in addition to the results: *Standard Search* and *Advanced Search*. The Standard Search merely enables you to enter a search term, as we did at the beginning of this example, so it is not particularly noteworthy. The Advanced Search can be very useful. On the Advanced Search tab, you can further refine your search by entering additional information about what you want to see and entering information about what you *do not* want to see. You can also control the number of links surfaced by the Advanced Search.

All of the search pages enable you to sort the result sets by ascending or descending date. And, they provide a way to hide the result summaries, so only the links are visible. Both of these features are useful ways for you to design the particular search that works the best for you.

- **Other Resources** – The bottom of the SAS Documentation home page contains links to two resources you might want to access:
  - *SAS Procedures by Name and Product* 9.3 | 9.2
  - *SAS Language Elements by Name, Product, and Category* 9.3 | 9.2

Both of these documents are great sources of basic SAS information.

SAS Help Menus in SAS Display Manager

Many SAS programmers don't seem to realize that there is a wealth of SAS documentation in the Help menus of SAS Display Manager. But, of course, being a top SAS programmer, you know they are there! This section provides an overview you can read in case you are not familiar with the many facets of the SAS Help menus.

The SAS Help menus are available via the Help selection on the top toolbar of SAS Display Manager.

Figure 5.2: SAS Help in SAS Display Manager

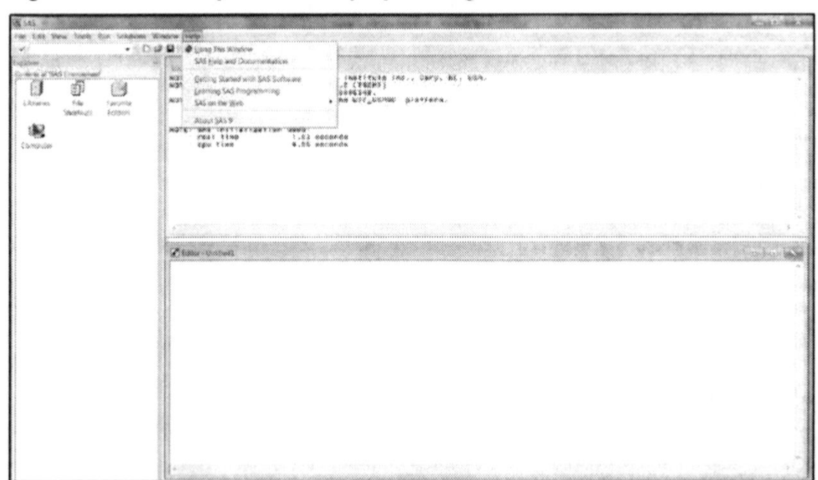

Six possible selections are available from the Help menu. The following sections describe each of them.

Using This Window

This selection provides a tutorial for using the SAS window, which is currently the focus of your SAS Display Manager session. For example, if you are currently in the Editor window, clicking **Using This Window** results in a tutorial on the Editor window. The same is true for the Log, Output, Explorer, and Results windows.

When you have an opportunity, take the time to browse through these tutorials and make sure that you are familiar with the details of using each SAS Display Manager window. You can also refer colleagues who are less familiar with SAS to these particular Help menus, so they can become more effective users of SAS Display Manager.

SAS Help and Documentation

This choice produces a very powerful SAS Help window you can use in a variety of ways to research SAS features and uses. The window has four tabs on the left vertical section of the window that facilitate research:

1. **Contents tab**– The Contents tab produces a tree view of four selections you can click on to obtain SAS information:
    - **What's New in SAS**– Provides information about what is new in the release of SAS you are currently using. Clicking on this selection produces a drop-down list of dozens of "What's New" topics, arranged by SAS product.
    - **Learning to Use SAS** – This topic produces a drop-down list with the following selections:

   - **Sample SAS Programs** – There are sample SAS programs for each SAS product you have licensed. These sample programs are a great way to learn about SAS software modules you do not have much experience using. When you click on an example, you can copy and paste the code from the Help window into the Editor window and then run it to see how it works.
   - **SAS e-Learning** – This choice provides you with a brief overview of SAS e-Learning classes. Self-paced e-Learning classes are available via the Internet to individuals and to organizations. They provide a convenient facility for learning new things about SAS.
   - **Tutorial: Getting Started with SAS Software** – This is a basic tutorial that provides SAS neophytes with the framework for using SAS. It is another good facility that you can bring to the attention of less experienced SAS programming colleagues in your organization.
   - **Accessing Help from a Command Line** – This pick surfaces links to descriptions on how to use the SAS Display Manager's Help facilities.
  - **Using SAS Software in Your Operating Environment** – This option reveals a drop-down list of SAS Companion documents, such as *SAS Companion for Windows*. It is handy to have such documentation at your fingertips within a SAS Display Manager session.
  - **SAS Products** – This choice produces a drop-down list of documents on topics such as Base SAS, SAS/GRAPH, and SAS/STAT. You point and click your way through the list in the left window. When you click on a document page, it is displayed in the window on the right, where you can read the documentation.
2. **Index tab** – This tab reveals an alphabetized list of SAS keywords as well as a keyword search window. As you enter your keyword, the long list of keywords moves down to match the alphabetic order of what you are entering. So, if you were to start entering "IBUFNO", by the time you entered "IBU", the keyword tree would position itself at "IBUFNO= system option." Simply clicking on the entry in the tree or clicking **Display** at the bottom of the page produces the text in the window on the right.

   The Index tab is obviously a useful tool for looking up SAS keywords and programming constructs. But you can also use it to simply browse and see if you can learn something new about SAS programming. There must be something somewhere in the index that you don't already know!
3. **Search tab** – The Search tab enables you to search the SAS Help documentation for specific keywords. You type the keyword in the search field and press **Enter**. Topics related to your search term are listed in the window. When you click on a topic, the text is revealed in the window on the right.
4. **Favorites tab** – When you find specific documents you are likely to be using a lot, you can add them to your Favorites tab window. Thereafter, they are readily available, so you don't have to search for them again. Whenever you have documentation on the Contents tab, Index tab, or Search tab, clicking on the Favorites tab results in the topic being placed in the "Current Topic" window at the bottom of the Favorites tab. Clicking **Add** then adds that topic to your favorites.

Getting Started with SAS Software

As the name implies, this main Help menu selection provides an entry to basic information about how SAS works. You are presented with two choices:

1. New SAS Programmer (quick-start guide) – This is a series of tutorials designed to get new SAS users who have a programming background up and running with SAS as quickly as possible. Share it with coworkers who are relatively new to SAS.
2. Experienced SAS Programmer (resource guide) – These tutorial pages are designed for experienced SAS programmers who want to know what's new in SAS and where to go for more support. You might learn something new in these pages.

Learning SAS Programming

This Help option facilitates the creation of sample data for SAS online training. If you license any of the SAS online training courses, clicking this link, and then clicking **OK** produces the sample data sets needed by the online SAS training classes. After the data sets are created, it makes the online training home page available. You can use that page to begin the particular online training classes you have licensed.

SAS on the Web

This Help selection opens your web browser to any one of five important SAS web pages. They are all great starting points for obtaining additional information about SAS.

- Technical Support
- Training Services
- Frequently Asked Questions
- Send Feedback
- Customer Support Center
- SAS Institute Home Page

About SAS 9

The final Help option opens a pop-up window that specifies information about the release of SAS currently installed. This is excellent information to have when communicating with SAS Technical Support or when determining the SAS tools available to you in the release you have.

SAS Publications

The SAS bookstore at http://support.sas.com/publishing/index.html offers a wide variety of hard-copy SAS documentation. Publications range from SAS technical references, to SAS course notes, to SAS Press books. For example, here are a few of the current titles for SAS/GRAPH:

- *SAS/GRAPH 9.3: Reference, Third Edition* – Technical reference
- SAS/GRAPH : Essentials Course Notes – Course notes
- *Quick Results with SAS/GRAPH Software* – SAS Press book

Technical references tend to be more formal descriptions of SAS software language constructs, configuration, and usage. They are written by SAS staff who are experts in the particular facet of SAS they are writing about. The course notes are exactly what they sound like: notes from SAS classes. You get the hard-copy course notes when you take a class from SAS Education. Or, you can order them from the SAS website. The course notes are generally very focused, step-by-step guides to using a particular product or feature of SAS. Like the technical references, they are also written by SAS staff who have expertise in a particular facet of SAS.

SAS Press books are written by SAS subject matter experts mainly in the SAS user community, although some are written by SAS staff. These books are expert guides to a particular component of SAS software. SAS Press books go through a rigorous technical editing process, so they are reliable references. The books tend to have a less formal tone to them, and generally fill in any knowledge gaps between SAS technical references and SAS course notes.

The Bookstore on the SAS website (http://support.sas.com/publishing/index.html) is your one-stop shopping spot for SAS publications. Here is an example of the home page:

Figure 5.3: Example of SAS Bookstore Page

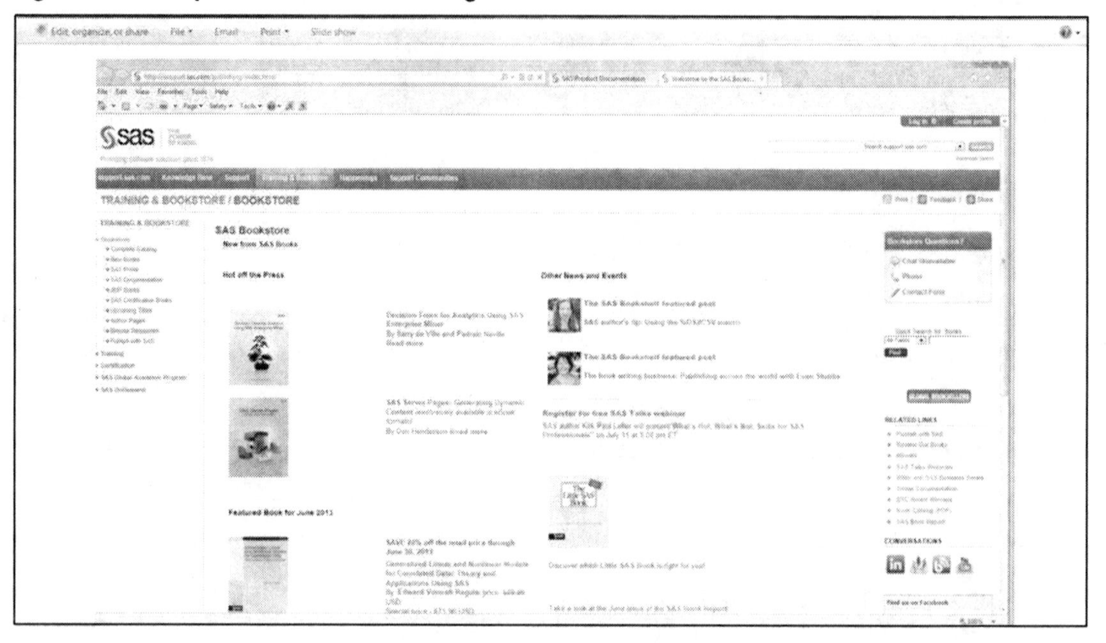

The SAS Bookstore home page provides a tree-view of 10 choices designed to get you to the SAS documentation you need:

- **Complete Catalog** –The complete catalog of SAS publications on a single web page.
- **New Books** – Newly published SAS books.
- **SAS Press** – The SAS Press books.
- **SAS Documentation** – SAS technical publications and course notes.
- **JMP Books** – Only books on JMP, SAS' premier statistical discovery software.
- **SAS Certification Books** – The preparation guides for those studying for SAS certification exams.
- **Upcoming Titles** – A list of upcoming titles soon to be published by SAS Press.
- **Author Pages** – SAS Press author biographies, their SAS publications, and more.
- **Browse Resources** – A page with general links to many SAS publication web pages.
- **Publish with SAS** – Information about how to write a book for SAS Press.

Being a top SAS programmer, you should have a library of hard-cover SAS publications you can quickly access to either study the particular features of SAS, or to reference when researching a SAS issue or question. With hundreds of SAS publications available, it might be difficult to know how many or which publications to purchase. All of them are informative and well-written. A number of the SAS Press books overlap in the subject matter that they convey. So, building the SAS library that is right for you might seem a bit daunting.

Worry not! The author has a suggested list of books you should consider for your personal SAS library. It is somewhat subjective, but should serve as a good starting point for your own deliberations on what works for you. A basic SAS library should have the following publications:

- **SAS Companion** (for your own operating system))
- **The Little SAS Book: A Primer, Fifth Edition,** by Lora Delwiche and Susan Slaughter
- **Learning SAS by Example: A Programmer's Guide,** by Ron Cody
- **Output Delivery System: The Basics and Beyond,** by Lauren Haworth, Cynthia L. Zender, and Michele Burlew
- **Carpenter's Complete Guide to the SAS Macro Language, Second Edition,** by Art Carpenter
- **PROC SQL; Beyond the Basics Using SAS,** by Kirk Lafler
- **PROC TABULATE by Example,** by Lauren Haworth
- **Cody's Data Cleaning Techniques Using SAS, Second Edition,** by Ron Cody

These books cover a wide variety of solid, hard-core SAS programming constructs and techniques. The information in them should encompass most of the types of programming tasks you perform in your SAS programming assignments.

If you have special needs, say you are working with SAS on z/OS, then you would consider targeted publications such as *Tuning SAS Applications in the OS/390 and z/OS Environments*, by Michael A. Raithel. Or, if you are doing factor analysis, you might purchase *A Step-by-Step Approach to Using SAS for Factor Analysis and Structural Equation Modeling*, by Larry Hatcher. So, become familiar with the entire SAS publications catalog to see if there is a particular publication suitable for the types of SAS programming tasks you most frequently perform.

Make sure you read the SAS publications that you purchase. Don't be a collector who simply lines a bookshelf with impressive SAS publication titles. Nobody is going to give you brownie points for having a great collection of books gathering dust on your bookshelf. There is a lot of good information stored between the front and back cover of each SAS publication just waiting for you to find and use. Read the books. Absorb the material. Exploit the knowledge found in SAS publications in your assignments so that you impress your colleagues and supervisors with your superior knowledge of SAS. That is one of the secrets to being a top SAS programmer.

Chapter 6: What You Can Do in Your Own Organization

Introduction

The phrase "think globally, act locally" is apropos when contemplating practical ways of becoming a top SAS programmer. By far, the best place to hone your SAS programming skills is inside the organization you currently work for. That is where you have access to SAS software on one or more computing platforms. That is where data exists in varying file formats and in database management systems. That is where you are asked to solve complicated business problems on a daily basis using SAS. That is where you collaborate with other SAS programming professionals on a variety of business tasks. Your workplace is where you can be most effective at both building your own reputation as a top SAS programmer and contributing to the good of your organization.

We are not talking about simply doing your job and completing the programming tasks assigned to you. It is a given that you do your job and you do it well. You already bring all of your computer programming skills and SAS knowledge to bear on your programming assignments. That is what you are being paid for and what you have been doing for years with your current employer and with past employers. We are talking about going beyond the common stimulus and response regimen of simply getting a programming assignment and completing it, getting the next assignment and completing it, and so on. You need to think about additional tasks that either increase your SAS acumen or contribute to strengthening the support of SAS in your organization. Such tasks benefit you, your colleagues, and your organization.

You are not going to be the top SAS programmer in your organization if you only do your job. That is no way to differentiate yourself from the other talented SAS programmers you work with. Your contributions must be larger than life. They should go beyond the normal bounds of project programming and encompass the SAS infrastructure of your company. You can do anything with SAS, so have confidence, and begin creating a track record of great accomplishments for yourself in your current environment. By making SAS resources more accessible to the other programming staff, you increase your value to your organization and gain greater visibility as the top SAS programmer. And that is exactly what you want to do, right? Right!

This chapter provides over a dozen ideas you can use to work at becoming the top SAS programmer in your organization. These ideas are designed to do three things: increase your SAS knowledge, increase your SAS reach, and increase your value to your organization. When you increase your SAS knowledge, you increase your effectiveness as a SAS programmer. Doing so can lead to more challenging assignments, greater accomplishments, and promotions. Increasing your SAS reach means doing additional things with SAS that are not currently being done in your company. They could be tasks designed to strengthen the overall SAS infrastructure and environment, or tasks done with SAS for other units within your organization. This widens the scope of the SAS culture and helps your company get increased value out of the SAS products it licenses. You increase your value to the organization by helping to facilitate other SAS programmers to become more effectual in using SAS and corporate SAS facilities. When other SAS programmers are more effective, assignments get done quicker and clients are happy. As the saying goes: *A rising tide lifts all boats*!

Study Other People's SAS Programs

One of the easiest ways to raise your SAS acumen is to study other people's SAS programs. Other programmers in your organization are likely performing a variety of interesting tasks with SAS. They are extracting data from relational database tables, creating custom reports with the SAS Output Delivery System, computing complicated statistics with SAS statistical procedures, cleaning dirty data, linking data from disparate file types, creating publication-ready charts and graphs, and doing a host of other noteworthy things with SAS. You might be engaged in some of these activities yourself, but probably not all. So, reviewing your colleagues' programs can give you a great education into what is going on in your organization in terms of data processing. And, it can expose you to a lot of different programming practices and methodologies. This can be helpful no matter what your level of SAS expertise is.

Other programmers bring a variety of programming experience levels and educational backgrounds to bear on their work. Your colleagues might range from graduates with Computer Science degrees to English majors to holders of associate degrees. They might have a couple of years of programming experience or have been programming for over twenty years. Perhaps they have worked for your organization all of their careers or maybe they have had a variety of previous programming jobs. They might have worked in a number of disciplines, or only in the one your organization is engaged in. All of these factors result in each programmer having a unique way of approaching assignments, and consequently, a unique way of writing SAS programs.

Reviewing your colleagues' programs can provide you with programming approaches and techniques that you might not have thought of before. It might supply you with ideas for improving your own SAS programming techniques. Other people's programs might contain SAS products, statements, functions, procedures, or options you are currently unfamiliar with or that you seldom use. Such programs could also provide insights into your corporate infrastructure, such as revealing the data sources available to you. You can use these real-life programs as examples of SAS programs that actually work in your computing environment and learn from them.

It shouldn't be hard to find SAS programs you can study within your organization. Look in corporate production libraries and public directories. Ask people for copies of their old programs. Examine the programs of other people working on your projects. Associates seldom mind sharing their programs when they find out that you intend to study them to learn more about SAS programming. It is this type of collaboration among workmates that makes an organization strong.

So, what should you do when you gain access to other people's programs? First, make sure you have the time to sit back and concentrate on just reading and absorbing the program. You want to have the time to understand the flow of the program. You need to realize its particular track on inputting data, processing data, and creating result sets. Here are a number of questions you should ask yourself as you read through the program:

- Does the program flow in a logical manner?
- Can you follow it?
- Does it use meaningful variable names?

- Is the program well documented?
- How easy is it to maintain and modify?
- Does it present new techniques you can incorporate into your own programs?
- Does it use SAS language constructs (procedures, statements, options, and so on) that you are unfamiliar with?
- Does the program work well? If so, why?
- Is there anything wrong with the program? If so, what?
- Can you make the program more efficient by using fewer DATA steps, PROC steps, observations, or variables?

Studying real-life examples of coworkers' programs is a great way to raise your SAS programming expertise. You can learn from their varied approaches to processing data, their different uses of SAS products and programming constructs, and their depth of understanding of your organization's requirements. And, while you are busy studying your teammates' SAS programs, make yours available for them to study, too.

Get a SAS Mentor

A SAS mentor is a person who has more SAS knowledge and experience than you have who agrees to counsel or guide you with your SAS programming. If you are relatively new to SAS, a mentor can be an invaluable resource for advancing your SAS education and for informing you of your organization's specific business practices. More experienced SAS programmers can also benefit from a mentor who can show them sophisticated SAS programming techniques and advise them how to best use them in ongoing programming assignments. Unless you are the top SAS programmer in your organization, having a mentor can enhance your programming career. So, take advantage of the SAS expertise in your organization.

Your prospective SAS mentors are the high-achieving SAS professionals within your establishment. They might be project managers, lead programmers, or prominent programmers and analysts. More than likely, they are currently considered to be some of the top SAS programmers in your organization. Currently! Whatever the case, it should not be too hard to identify the small group of high-value SAS professionals in your business who are your pool of prospective mentors. Once you identify them, you need to winnow down the choices to find the one who is right for you. Issues such as personality, access to the individual, and the person's probable willingness to participate are some of the things you should consider when determining who to contact.

Approach your potential mentor and explain that you are relatively new to SAS programming, or conversely, that you are very experienced but would like to learn more. Acknowledge the person's advanced command of SAS programming. Ask if the person would be willing to answer a few questions occasionally about SAS programming constructs and techniques. Is the person agreeable to providing insights into SAS programs you are working on and data sources within your organization? Would your prospective mentor teach you new SAS programming techniques and review how you use them in your programs? Can you occasionally ask this person how to approach a particular programming assignment? Would it be possible to review some of the mentor's old SAS programs to learn new techniques? Can that person recommend particular SAS publications that would enhance your knowledge of SAS programming? If you have done your homework correctly and approached the right individual, more than likely the person will say yes and you will be on your way to having a mentor.

Once you have secured a mentor, take advantage of the person's willingness to help. Ask pertinent questions about your SAS programs, data sources, SAS constructs, programming techniques, and corporate data sources. Bounce ideas and possible programming approaches off your mentor. But, be careful not to wear your mentor out. Remember that this individual also has programming assignments to carry out, and might also have management duties to attend to as well. So, be sure to strike a balance so that your mentor does not feel like you are unduly leaning on her to get your work done. Doing so will benefit your relationship with your mentor and ensure that your SAS programming expertise increases under the mentor's tutelage.

Learn New SAS Programming Techniques

Because the SAS programming language is so vast, it offers almost unlimited opportunities for you to learn new programming techniques. There are hundreds of procedures, statements, functions, options, formats, and so on, available for you to master. They perform operations as varied as accessing a host of file types, inputting data, deriving new variables, computing advanced statistics, and creating sophisticated reports. Not only do you need to learn the syntax of such language constructs, but you need to learn the intricacies of actually using them to input, process, and output data. So, there is *always* something new for you to learn in the world of SAS programming.

When you think about it, this is what you should be doing, anyway. Right? To become a top SAS programmer, you must always be looking for new ways of using SAS to perform your job functions. You need to master new techniques to continue to evolve and get better and better at programming. You have no choice in the matter; this is something you must actively do. Learning new SAS programming techniques keeps you on the cutting edge of SAS programming.

Browse through the books in your personal SAS library and examine the SAS Online Documentation looking for procedures, coding techniques, tips, and tricks that you do not already know. Experiment with them to see how they can enhance your current assignments. For example, can you do the next assignment using PROC SQL instead of using DATA and PROC steps? Can you use SAS macros to reduce redundant code in your programs? Can you harness the Dictionary Tables and SQL Views to let the metadata drive a particular program? Are there facilities of the Output Delivery System that you can harness to create more sophisticated and professional-looking reports? Could you automate your SAS programs by using .bat files and the job scheduler on Windows, or scripts and the cron tab on UNIX? How about reworking your programs so that the reports are e-mailed to users? There are hundreds of possible programming features and techniques that you can bring to bear in your current SAS coding assignments. So, why not learn something new?

Here are other sources you can plumb for new SAS programming ideas:

- the discussion forums on support.sas.com
- samples and usage notes on support.sas.com
- SAS users group conference proceedings
- your colleagues' programs
- the SAS-L listserv

Because there are always new programming techniques available, you should never get bored with SAS. Make sure that you are on the cutting edge of SAS programming by staying up on the current trends and ideas in programming.

Become Your Organization's Liaison with SAS Technical Support

How does your organization handle SAS technical support questions? In some organizations, it's "every man for himself" and each SAS programmer contacts SAS Technical Support individually for technical problems. This is a wasteful approach for several reasons. First, it does not facilitate keeping a corporate history of known issues and their solutions. Programmers resolve their individual issues and go on their way, while others might trip over the same issues. Secondly, many programmers do not know how to properly document and report an issue to SAS Technical Support. Although the support staff will lead them through the necessary steps, programmers lose valuable time stumbling through the process. Thirdly, some issues can be resolved by knowledgeable in-house staff members. So, programmers waste valuable time documenting and reporting problems that could be quickly resolved by using advice from their colleagues. Finally, there is not an easy way to determine that a particular problem is emerging as a systemic issue within the organization's SAS infrastructure and needs a global fix. So, many staff members might trip over the problem individually before somebody finally realizes that a universal solution is needed.

A centralized liaison with SAS Technical Support provides many benefits to an organization. Here are some of them:

- The liaison can triage SAS technical issues before they are reported to tech support. Some questions are relatively easy and can be answered right away by a knowledgeable SAS professional such as yourself. Other issues might have previously been submitted to SAS Technical Support, so the answers are already known and can be communicated directly to users. This obviates the need to communicate with SAS Technical Support and saves everybody time.
- The liaison can help the user define the specific question, gather the necessary documentation (programs, logs, and maybe data), and report the incident to SAS Technical Support. The liaison can follow up with the user and technical support when additional information is needed. The liaison can communicate the answer directly back to the originator, putting it in terms the user can understand. This protocol frees staff members so that they can engage in project work that adds value to the organization while the problem is being researched.

- The liaison can document issues that might affect lots of SAS programmers in a public forum for the organization, such as a Microsoft Outlook information folder or a "SAS Issues" page on the corporate intranet. This benefits all programming staff by providing them with an in-house resource they can access to determine whether there is a solution or workaround to a problem they are experiencing.
- The liaison can determine when a pattern of problems related to the same issue is emerging. Consequently, the liaison can communicate with a larger group of staff members that might be affected by the issue and provide a proactive fix.

So, you need to become your organization's liaison with SAS Technical Support. This is a great way for you to stay up on what is happening with SAS programming in your organization, on the current thinking about SAS code, and current SAS technical problems. It is also a good way to get to know the friendly, helpful folks of SAS Technical Support. But, where do you start?

SAS encourages organizations to designate on-site personnel in the following positions:

- **The SAS Installation Representative** The person who receives the SAS installation media (tapes, disks, or CD-ROMs).
- **The SAS Training Coordinator** The contact person for SAS training (computer-based, video, and instructor-based) at your site.
- **The SAS Contract Representative** The person who receives contractual documents.
- **The Invoice Contact** The person or department that received invoices.

Your organization might already have staff who are in contact with SAS for various matters. The person who is likely the most conversant with SAS on support issues would be the SAS installation representative. That person might or might not provide a central source of contact for issues reported to SAS. If not, then you have an opportunity to provide a service that will help your organization. If so, approach that person and see if it is feasible for you to undertake the *burden* of being the liaison with SAS.

Work with management to have all SAS support questions routed through you. This requires you to build a bit of infrastructure:

- **E-mail address** Consider acquiring a unique e-mail address coworkers can use to route all SAS technical questions to you (for example, SASsupport@yourorganization.com).
- **Phone number** Publish your phone number as the SAS technical support contact on the corporate intranet so that staff members know who to call.
- **Outlook information folder** Create a SAS Outlook information folder that staff can use to pose technical questions, and that you can use to post answers as well as SAS issues of general interest.
- **SAS page on corporate intranet** Work with your webmaster to establish a SAS page on the corporate intranet. There, you can advertise your contact information and post material of interest to SAS users in your organization.

- **Advertising** Look for opportunities to advertise that you are the go-to person for technical support. This includes spots in corporate newsletters, e-mails to SAS users, contact information posted on the corporate intranet, perhaps even a note in SAS News in the SAS log.

The key to being successful in the liaison role is to make sure the programming staff gets satisfactory answers to each issue they report in a timely fashion. You must work closely with them to frame the question and get the background material. You need to report the problem to SAS Technical Support and work closely with them on follow-up questions. Then, you should frame the answer in such a way that the originator understands it and is able to move forward with the SAS task at hand. This raises your corporate visibility and provides a beneficial service to your organization.

Organize SAS Program Code Reviews

A *code review* is exactly what it sounds like: an event in which a group of programmers gather to review the code of one or more programs. Code reviews are usually held as a last quality check for programs that are going to be put into production mode to ensure they perform the tasks they are supposed to perform. Participants follow the flow of the program, match the program code against the specs, validate the program's methodology, certify the program's output, and provide the author with insights into more efficient coding practices. The result of a code review is a program that has been validated to perform the functions detailed in the specifications in the most expeditious manner. A more subtle result is that all people involved gain a better knowledge of the application, the data, and group programming practices.

Although code reviews were once a mainstay of creating production systems, they seem to have fallen out of favor and are not embraced by many organizations today. Implementing them can provide you with a good opportunity to contribute to your organization while learning new things about your group's applications and individual SAS programming style. So, think about how you can set up code reviews for your coworkers and yourself for SAS programs that are slated to be put into production.

This is a general protocol for a code review:

1. Talk to your colleagues and agree on one or more programs to review. It is a good idea to review several people's programs in a single sitting so that nobody feels like they are being singled out for criticism.
2. Schedule a meeting room and invite the programming team. Make sure you allot enough time, because technical discussions can sometimes be lengthy. Let the team know whose programs are going to be reviewed.
3. Each person who has a program to be reviewed brings handouts:
    - the SAS program
    - the specifications for the assignment

   - the SAS log from running the program
   - output of the program – reports, PROC CONTENTS of SAS data sets, layout of flat files, and so on
4. One by one, each individual walks the group through their program, starting at the top and working down to the end. They describe what is being done by the various DATA steps and PROC steps, and the context for the logic by referring to the specifications. They show how the report or output file was derived and how it conforms to the requirements.
5. The group provides feedback by noting possible flaws in the logic or anomalies in the SAS program. The group actively asks questions and seeks clarifications. When appropriate, group members make suggestions for tightening up the program's code, or for using more efficient programming techniques, or for finding alternate ways of achieving the same results. Here are some things to consider:
   - Does the program flow in a logical manner?
   - Can you follow it?
   - Does it use meaningful variable names?
   - Is the program well documented?
   - How easy is it to maintain and modify?
   - Does the program do what the specifications require it to do?
   - Is there anything wrong with the program? If so, what?
   - Can the program be made more efficient by using fewer DATA steps, PROC steps, observations, or variables?

It is very important that all people involved undertake a constructive, professional approach when participating in a code review. Remember that it is easy for criticism to sound biting if not delivered in the right tone. You don't want your colleagues to feel defensive in a code review. Rather, you want them to feel collaborative and empowered. So, pay careful attention to the atmosphere in the SAS code review and proactively rein in discussions that do not appear to be constructive. Doing so can help your colleagues look forward to sharing their programs and seeking advice from the group in future code reviews.

You should be the person who sets up the SAS code reviews. Keep an eye on which programs are coming up to be scheduled into production. Make contact with your colleagues, have them agree to the review, schedule the room, and invite the right people. Publish an agenda of the order in which your colleagues will present their programs. Then, act as the moderator. Make sure each person has a chance to present their programs. Keep the discussions on track and on time. Make it a point to intercede in those rare instances when the criticism doesn't appear to be well taken by moving the discussions on to other topics.

SAS code reviews can strengthen your own programs and your coworkers' programs. You can all learn from each other. You learn new SAS programming techniques, new approaches to business problems, and the nuances of your organization's data. Code reviews can also strengthen the sense of camaraderie you and your workmates have with one another. And, all of this is happening while you put well-reviewed SAS programs into production.

Make SAS More Efficient in Your Organization

A fairly easy way to make a good name for yourself while benefiting your enterprise is to make SAS more efficient in your organization. It is *easy* because many people do not take the time to make sure their SAS programs are written as efficiently as possible. But, it is not really their fault. Most of your colleagues are probably writing programs under deadline pressure. They are more concerned with writing programs that match specifications and produce the desired outputs on time than with how efficient their programs are. So, many application programs are written in haste with the end product—not efficiency—in mind. This results in a treasure trove of SAS programs in production libraries that can be fine-tuned to be more efficient.

If you have the time to invest, this phenomenon presents you with a great opportunity. Coming along after the fact, and with no deadline pressures, you have the luxury of reviewing existing programs that are still in operation for ways the code can be tightened up. You can determine ways in which production programs can be more efficient and either modify them yourself, or recommend changes to the authors. You can give rise to and drive a process that nobody else is thinking about: the process of making SAS more efficient in your organization. But what does "efficient" mean in this context?

A program is efficient if it uses the minimum amount of computer resources possible to complete its task. Computer resources are factors such as computer memory, Input/Output events, user CPU time, system CPU time, disk storage, and network bandwidth. Reducing these factors leads to reduced overhead on your computer systems and faster execution times of your programs. In organizations that have chargeback systems for elements such as CPU, I/O events, and storage, you can save your group money. You might even be able to help the entire organization by reducing the amount of data transferred over the network, thus reducing network bandwidth. These are all win-win factors for you and for your teammates.

The SAS log provides some elementary metrics you can use to determine program efficiency. Some other sources are the FULLSTIMER option that prints all available system performance statistics to the SAS log, the LOGPARSE macro that enables you to store SAS performance statistics in a SAS data set for later reporting, the SAS interface to Application Response Measurement (ARM) which enables you to record a SAS program's performance statistics, and the SAS SMF exit which records performance statistics in the z/OS environment. You should get familiar with these vehicles for measuring program performance and with the many metrics they measure.

You can find methods for making SAS run faster and more efficiently in the discussion forums on support.sas.com, samples and usage notes on support.sas.com, SAS users group conference proceedings, your colleagues' programs, and the SAS-L listserv.

Actively seek out and research SAS performance tips from those sources and take note of them. Experiment with SAS efficiency techniques. Can you make SAS run faster? Can you reduce the cost of running large SAS programs? Can you decrease the overall size of the SAS data sets stored on network directories? Can you improve on the SAS option settings found in the autoexec.sas and sasv9.cfg files? If you can do any of these things, make recommendations for your team and to the SAS installation representative. Show them the proof of what you have discovered and help them understand the positive ramifications of rolling your proposed changes out to the team or to the entire organization. Then, push to have those changes made.

It should be easy enough to find programs that are candidates for being made more efficient. Comb through production program libraries and look for likely suspects. Larger, more complicated programs are likely to yield more tuning opportunities. Be systematic and keep track of the programs you are looking at. Once you have some ideas on how they might be modified, contact the authors and discuss your ideas. Be careful to let the authors know you are not critiquing their programs, but simply suggesting ways the programs can be made more efficient. Hopefully, they will agree with your assessment and then either modify the programs or let you modify them. After a program has been modified, compare the new SAS log with the old one and determine how much computing overhead was saved by the changes.

Make sure you publicize all of the positive recommendations you made and all of the resources and cost reductions you were able to reap. Bring the savings to your supervisor's attention and your coworkers' attention. It is true that you are making SAS applications more efficient primarily to benefit your organization. However, you should get credit for taking the initiative, doing the work, and making a measurable difference in your enterprise.

Start an In-House SAS Users Group

You can exhibit leadership and become known as the top SAS programmer in your company by starting an in-house SAS users group. An *in-house SAS users group* is simply a SAS support group hosted within your organization. SAS users groups usually hold meetings on a cyclic basis (for example, monthly, quarterly, biannually). Meetings are typically an hour long and open to all SAS users in the organization. SAS users group meetings generally feature a SAS technical presentation, a question and answer session on the topic presented, an open-forum SAS question and answer session, and news about corporate SAS initiatives. In-house SAS users group meetings provide great venues for SAS professionals to network with one another and trade SAS programming tips and techniques.

SAS provides support for in-house SAS users groups. All you need to do is register your group with SAS via a link on support.sas.com. (Select **Support & Training→Support Communities.)** Once your group is registered, SAS appoints an official liaison for you to work with. The liaison helps facilitate your group getting a SAS speaker and some great giveaway items for one of your meetings. The liaison can also apprise you of other educational opportunities for your meetings such as webcasts. So, be sure to register your in-house SAS users group with SAS.

The SAS Users Groups web page lists some of the possible benefits of participating in a SAS users group:

- increased efficiency and productivity through increased exposure to the following:
    - new coding techniques
    - analysis techniques
    - SAS applications
- opportunities to polish your interpersonal, writing, presentation, and leadership skills
- enhanced understanding of SAS' software and service
- networking and idea sharing with other SAS software professionals

All of this is very true both for you and for your colleagues. In addition, the users group provides a venue in which you can all share SAS knowledge uniquely pertinent to your organization such as when you will be upgrading to the next release of SAS.

To start an in-house SAS users group, first make sure you have your management's support. Then, register your group with SAS. Next, work with your Customer Account Executive (CAE) to create a survey that will help determine logistics of the meeting; that is, how often you will meet, what time of day is best for your peers, what topics are most popular, and so on. This survey will help you get the feedback you need from your SAS colleagues to set up your group in a way that best fits their needs. One big factor in how often you meet is your determination on whether you can line up a presentation for each meeting. It might be best to start off with quarterly meetings and then add more if there is a great interest on the staff's part in giving presentations. You also need a way to advertise the meetings to your organization's community of SAS users. E-mail is the most convenient way to make staff aware of upcoming meetings. Your CAE and users group liaison can help create meeting invitations if you want. Your in-house SAS administrator might be the best source for your building an e-mail address list of perspective attendees.

Once you have an e-mail list and a presenter lined up, you can schedule your first meeting and send e-mail announcements to your SAS user community. It is a good idea to send an initial announcement the week before, and then a follow-up announcement the morning of the presentation. That puts the meeting on people's mental calendars and then provides a reminder as they contemplate their busy schedules on the day of the meeting.

Here is an example of an invitation:

All,

This Friday, join Michael A. Raithel as he presents ***PROC DATASETS: The Swiss Army Knife of SAS Procedures***. This tutorial highlights the versatility of the DATASETS procedure, which can be used for everything from deleting, copying, or renaming SAS files, to adding labels and formats to SAS variables, to creating and controlling audit trails, to repairing damaged SAS data sets. With over two dozen statements and several score of options, PROC DATASETS provides a function-rich utility for managing SAS files. Come and learn a new use or two for PROC DATASETS that you can apply in your own SAS programs.

Friday's presentation is:

- Date/Time: **Friday, April 27, 12:00 - 1:00**
- Place: **Conference Room 0427**
- Presenter: **Michael A. Raithel**
- Topic: **PROC DATASETS; The Swiss Army knife of SAS Procedures**
- Abstract:

  The DATASETS procedure provides the most diverse selection of capabilities and features of any of the SAS procedures. It is the prime tool that programmers can use to manage SAS data sets, indexes, catalogs, etc. Many SAS programmers are only familiar with a few of PROC DATASETS's many capabilities. Most often, they only use the data set for updating, deleting, and renaming capabilities. However, there are many more features and uses that should be in a SAS programmer's toolkit.

  This paper highlights many of the major capabilities of PROC DATASETS. It discusses how it can be used as a tool to update variable information in a SAS data set; provide information on data set and catalog contents; delete data sets, catalogs, and indexes; repair damaged SAS data sets; rename files; create and manage audit trails; add, delete, and modify passwords; add and delete integrity constraints; and more. The paper contains examples of the various uses of PROC DATASETS that programmers can cut and paste into their own programs as a starting point. After reading this paper, a SAS programmer will have practical knowledge of the many different facets of this important SAS procedure.

Come early so that you can get a good seat!

Sincerely,

Michael A. Raithel

This is an example agenda for your meeting:

- Welcome everybody to the meeting.
- Announcements of future SAS users group meetings, including the date, the proposed speaker, and the topic.
- Discussion by the SAS installation representative about relevant SAS initiatives such as migrating to a new release.

- Introduction of the presenter (state the name of the presentation and provide a brief, one-paragraph biography of the presenter).
- The SAS technical presentation.
- Questions about the presentation.
- A general SAS question and answer session. Anybody can ask a SAS question of the group and anybody who can provide an answer, or an idea on a possible direction to take, replies.
- A reminder of the date of the next meeting and closing of the current meeting.

One of your biggest challenges is going to be finding speakers for your meetings. Encourage your colleagues to give presentations about some of the interesting things they are doing with SAS. They can simply be using SAS in a unique way, or in a way that others might benefit from knowing about. If it is okay with management, invite speakers from other organizations. If you comb through regional and local SAS users group conference proceedings, you can usually find SAS experts who live in your area. Contact them and invite them to give their conference presentations at one of your users group meetings. There are also organizations that have speaker-sharing programs, such as SAS Press and the SouthEast SAS Users Group (SESUG). Get in touch with them and see if they are willing to help you procure a speaker for one of your meetings. Finally, if you are located in a large metropolitan area, it might be possible to form a speaker-sharing arrangement with other organizations that have in-house SAS users groups. Contact your SAS users group liaison and ask her for the contact information for other in-house SAS users groups within your area.

Another idea for recruiting speakers is to hold a "Coders Corner" type SAS users group meeting. In a Coders Corner meeting, several participants give five-to-ten minute presentation on particular SAS topics of their choice. Topics can range from SAS procedures, to programming language constructs, to programming techniques, to how to process in-house data sources with SAS. It is usually much easier for people to put together a brief, targeted presentation than to craft an hour-long presentation. You can often find enough SAS professionals to fill an hour. All it takes is five ten-minute presentations—counting transition times between speakers—to fill an hour-long session. Your speakers will appreciate the opportunity to present, and the audience will appreciate the variety.

Consider being the presenter at your first in-house SAS users group meeting. It will help establish you as a SAS expert within your organization and display your SAS acumen to your peers. Whether you present or not, make sure you are the emcee and the meeting facilitator. You should be the one who calls the meeting to order, makes the announcements, introduces the speaker, moderates the Q&A, keeps track of the time, and so on. Doing so provides you with greater visibility among your SAS peers, which, of course, you deserve for organizing the group.

Become a Mentor in Your Organization

If you are well on your way to becoming a top SAS programmer, consider becoming a mentor to less experienced SAS professionals in your organization. That is a great way to give back to your company while raising your own corporate profile. Mentoring other programmers helps raise their overall knowledge of SAS and of SAS programming techniques, making them more effective in completing their assignments. It also reinforces your own knowledge of SAS as you address situational and detailed SAS programming questions you might not otherwise have run across. Your initiative in mentoring less experienced staff shows management that you have leadership qualities and are interested not only in your own job, but in the greater corporate interest. Mentoring is good for you, good for your coworkers, and good for your organization.

To get started, look around the office for novice SAS programmers who might benefit from some help. Approach them individually, as appropriate, and let them know you are available to answer SAS programming questions. Make sure they understand that you won't do their programming assignments for them, but will act as a sounding board and an advisor. (You need to set some limits so that you can get your own work done.) Advise your prospective mentees of the best way to contact you with an issue: e-mail, phone call, drop by in person, or all three. If they are relatively new to the organization or to SAS, make them aware of the various corporate facilities available for supporting SAS, such as a users group, lending library, or SAS web page on the company intranet.

You don't necessarily have to have a formal agreement where you and your colleagues officially designate each other as mentor and mentees. All it takes is an understanding that they can contact you with SAS programming questions when they are stuck and you will provide your best advice. You want to become known as *the SAS answer person* in your organization. You should be the person your mentees (and others) come to for SAS information and answers to SAS programming questions.

Once you have established your mentor relationships make sure you take an active role in keeping them alive. Check in on your constituents on a regular basis by dropping by and asking what they have been working on. Chances are good that they will tell you about any SAS programming problems they have been having or odd errors they have experienced. As they describe issues or show you their programs, be alert for tips you can offer about more efficient programming techniques they can use. Perhaps they can perform the task with fewer DATA or PROC steps, or "macrotize" a program, or use a different procedure or function to more efficiently process the data. Offer advice and suggestions in a collegial manner so that your mentees are more focused on what you are saying than how you are saying it. Overall, a proactive approach to mentoring keeps you accessible and in the loop with your constituents.

Establish an In-House SAS Publications Lending Library

An additional way to mentor your colleagues is to establish an in-house SAS publications lending library. The "library" might simply be a section of one of the bookshelves in your cubicle or office. You stock the library with key SAS Press publications. Staff can drop by and borrow books at their convenience after signing them out via a form on a clipboard. Hopefully, they will consult with you on the appropriateness and usability of a particular publication before checking it out. Be sure to actively advise them on the right book for the particular topic area they are interested in. Also, get the word out that the library exists and is available to all interested SAS users in your organization.

The key to a successful SAS publications lending library is to have a good selection of the right books available for borrowers. (See Chapter 5, "SAS Documentation" for some recommended titles.) Once you have built up a suitable library, let your colleagues know it is available. Send an e-mail to the SAS users in your organization announcing the SAS lending library and listing the available publications. Send a reminder e-mail once a quarter with the list of titles, highlighting any new publications that are available. Be sure to ask staff to send feedback on particular publications they are interested in having in the library.

Keep track of the books in the library and those that have been checked out. Periodically, visit your constituents and see if they are still using the particular publications they checked out or if they are done with them. Make sure that you subscribe to the SAS Publications catalog so that you receive periodic updates on available SAS books. If somebody drops by with a need for a specific book not already in the library, order it. Being proactive and responsive are key ingredients to having a dynamic SAS publications lending library that is actively used by your SAS programming associates.

Try Out the Latest SAS Products

You can raise your overall knowledge of SAS by getting your hands on the latest SAS software products and learning to use them to process your organization's data. First of all, go after the SAS products your organization has already licensed. Run a PROC SETINIT to determine all of the SAS modules currently in your SAS installation. Then, ask yourself if you have mastered all of them. If not, make it your goal to become proficient in each and every one of the SAS products your organization has. Refer to sources of information about SAS such as online documentation, SAS Press books, the SAS Sample Library, and SAS users group conference papers. Read, experiment, and learn until you become proficient with the suite of SAS your company has licensed.

Think of ways you might be able to try out SAS products your organization does not currently license. SAS occasionally offers introductory webinars about its software, so you can learn about it from the comfort of your own desk. Sometimes local SAS training centers offer one-day meetings where various SAS software is demonstrated. Experts are on hand to provide demonstrations, answer your questions, and lead you through using the software on test data. Regional and international SAS users group conferences usually feature a "demo room" where SAS programmers such as you can talk to specialists to learn about new SAS products, and to learn about new features of existing SAS products. Your hard-working SAS representative should be able to provide you with information about webinars, SAS demonstrations, and the like. She can also provide information that keeps you up-to-date on the latest SAS software offerings.

Take a good look at your company's business processes and consider how additional SAS software could make a positive impact. Which SAS software modules that could help your business have been overlooked? For example, maybe your organization should be using SAS on other platforms such as UNIX, Linux, z/OS, or Windows servers. The faster processors, larger memory, greater work areas, and easier access to production databases on those servers could lead to faster turnaround time for SAS programs. So, you would consider licensing Base SAS and other relevant SAS products for your servers. You would also want to license SAS/CONNECT software for both the workstations and the servers. SAS/CONNECT would enable you to easily transfer data between operating systems, and to submit SAS programs from PCs that run on the servers. Then, your organization's PC SAS users could submit programs that ran on the servers, accessed the corporate databases, produced analysis files, and downloaded those files to their workstations. Wouldn't that make a positive impact!

Make it a point to learn about the latest SAS products and to try them out whenever possible, wherever possible. Keep your organization's business and computer processes in mind and think of how new SAS products could make them more efficient. Build a business case for the SAS product you have in mind and bring it to your management's attention. Work with your management and with your SAS representative to order the software and then with your SAS installation representative to have it installed. Perform some preliminary tests to ensure you know how it works in your environment. Then, let your coworkers know the new product is available. Consider publishing a "how-to" guide to using the new SAS module, complete with links to the online SAS documentation. Finally, make sure your manager is aware of your initiative and positive contribution to the business processes of your organization.

Get to Know Your SAS Representative

One of the hardest working people you are ever likely to meet is your SAS Representative. She works with dozens of organizations on a myriad of SAS product licensing and support issues. Your SAS rep does everything from recommending particular SAS modules and SAS solutions, to furnishing SAS technical staff who can address your SAS problems, to providing pricing information, to helping to consummate licensing paperwork and getting you the SAS modules that you need. She is well-informed on SAS products and issues, and well-connected with SAS consultants who can provide software advice and technical support. Your SAS representative is constantly making the rounds of the various organizations that she works with to inform them of new SAS initiatives, and working to show them how they can better use SAS for a better return on investment. SAS representatives tend to be human dynamos.

Because your SAS representative is well-connected and genuinely helpful, you should take the time to get to know her. If you are not the SAS software representative in your organization, reach out to her by e-mail or phone and introduce yourself. Let her know that you are the top SAS programmer in your organization and are interested in keeping up with the latest information about SAS. Ask your representative if it would be possible to stop by and meet you the next time she is visiting your organization. If you have any SAS questions, ask her what she thinks might be the best channel of communication for addressing those questions. Make sure to coordinate your contact with your SAS software representative through your organization's SAS software representative so that you don't step on any toes.

Your SAS representative has deep connections within SAS. Don't struggle with questions you might have about SAS software modules, platforms SAS is supported on, or pricing; just ask her.

She can either answer such questions flat out or have other SAS technical staff contact you for a discussion. She can put you in touch with other organizations that license SAS and that have similar computing environments and software issues. She can help you with information about SAS classes, seminars, webinars, SAS classes, SAS e-learning, SAS users groups, and dozens of other SAS issues. So, take that first step and make contact. It will be the beginning of a mutually rewarding business relationship.

Create SAS Pages on the Corporate Intranet

A very public way to provide a service to your fellow SAS programmers is to create SAS web pages on the corporate intranet. The SAS web pages should provide information about using SAS in your organization and references to where staff can go to in order to find additional information about SAS. A good model is to have a SAS information home page with links to subordinate pages based on content. For example, you might have a sub-page for corporate SAS training, one for information about the current version of SAS installed, one for how to reach internal SAS technical support, and so on. The in-house SAS website should provide a *one-stop shopping* destination all SAS professionals in your organization can call upon to orient themselves to all things SAS in your organization. This is helpful to new SAS programmers and to programmers who are new to your organization who want to understand the SAS programming scene. Organizational SAS pages are also a great way to communicate the inevitable changes to the SAS computing environment as new releases are installed and new SAS products are purchased. They also raise the visibility of SAS as a power programming tool in your organization.

If you are thinking, "I am a SAS programmer, not a web developer; how would I pull this off?", don't despair. If you can obtain corporate support for the SAS pages and can get a web developer assigned to the task, then it can be pretty straightforward. Use your organization's word processing software to create the content for the web pages. Create the mock-up of the home page with links to the sub-pages. Then, create the content for each of the sub-pages, including relevant links to non-corporate SAS facilities. Pass these along to your talented web programming associate so that she can turn your prototype into a web application on your intranet. It can be as simple as that!

Providing easily accessible information about SAS can also be a proactive way of heading off technical support issues. Informed, educated SAS users are better consumers of corporate SAS resources and less likely to contact the technical support staff unless they have a significant issue. So, when you are plotting out what to post on the website, err on the side of putting *more* rather than *less* information. More is better! Here are some suggestions for pages to link to from the home page:

Classes Provide a list of upcoming in-house SAS classes, including dates, times, how to sign up for a class, class prerequisites, and class descriptions. Also post links to the SAS Support & Training page on support.sas.com so that staff can consider SAS live or e-learning classes.

Conference Papers List SAS users group conference papers your organization's staff have written with links to them.

Documentation Include a page of links to SAS online documentation for each of the SAS products your organization licenses. You could also include instructions on this page for how staff can order hard copies of SAS publications.

Fact Sheets Link to in-house-written papers concerning various SAS topics and techniques particular to your corporate environment. Papers might range from how to use SAS/CONNECT, to how to link to your organization's UNIX servers, to how to send e-mails via SAS, to how to find the best ways of accessing corporate databases.

Frequently Asked Questions Post answers to commonly posed questions regarding the installation and use of SAS software in your organization.

In-House SAS Technical Support List information about how to obtain in-house SAS technical support, including the e-mail address and extension of the support staff, and the hours of operation.

In-House Users Group Information Provide an overview of your in-house SAS users group and a schedule of upcoming presentations.

Links Include hyperlinks to a number SAS websites hosted by universities, SAS users groups, and private individuals who provide useful information about SAS programming.

SAS Products Licensed List all of the corporate SAS platforms and the SAS software products installed on each of them. Programmers can use this list to determine whether the SAS products they are interested in using are currently on the computing platforms they are using.

SAS Software Update Schedule Provide staff with the schedule for upgrading SAS to a new version or maintenance release, and also the schedule for when hot fixes are applied to SAS servers.

SAS Tips Compile simple, one-or-two paragraph tips that make SAS programming easier for your coworkers.

Version Information Include a page with information about the versions of SAS software your organization currently supports.

Though not exhaustive, this list of possible web pages should give you a starting point for creating your own in-house SAS links. Use your imagination and ask yourself what information you would want to have if you were new to SAS programming in your company. What things would you like to know about how SAS is installed, the computing platforms it is supported on, how technical support is handled, what SAS products are licensed, and what additional corporate SAS resources are available to you? The answers to these types of questions should drive the web pages you create for your constituents who use SAS.

Take care to review the contents of your corporate SAS web pages on a routine basis to ensure they are up-to-date. Few things turn intranet consumers off faster than finding out the information about a website is old and unreliable. For example, when you migrate from SAS 9.2 to SAS 9.3, review your web pages and confirm that you have updated them to reflect the change. If you have an FAQs page, check to verify that it is still relevant with the latest release of SAS. As SAS gets better and better, some old programming constructs and concerns are supplanted with new constructs within the programming language. You do not want to perpetuate old workarounds if newer SAS software addresses them. So, check your FAQs page regularly to make sure it is up-to-date.

Make certain your name and contact information are visible near the top of the SAS pages home page. Although you are providing an important public service to the SAS users in your community, your goals are not entirely altruistic. Your colleagues and management should know you are the answer person for general SAS questions and questions about corporate SAS resources. This tells them who to contact with feedback on the things you post to the website. Placing your name and contact information at the top of the page means that all users can see it and begin associating you as the go-to SAS person in your organization.

Get Published in the Corporate Newsletter

One of the biggest problems with a corporate newsletter—in fact, with any newsletter—is finding enough content to fill it. Newsletters seem like a good idea at the start, but after a few editions editors find themselves scrambling for articles. That is why organizational newsletters often degenerate down to one or two short articles of general interest and lots and lots of announcements of scheduled corporate events. The scheduled events are relatively easy to come by. Finding quality articles is hard.

You can take advantage of this fact by writing and submitting short articles to your corporate newsletter that are focused on SAS. Contact the editor and let her know you have an idea for an article about SAS that will be of interest to SAS programmers in the company. The more SAS users you have in your organization, the better your case for having SAS articles in the newsletter. So, if you know the number of SAS users in your company, let the editor know about her potential readership. However, if there are not a lot of SAS programmers, tell the editor your article will be educational and allow other users to gain insights into the amazing world of SAS programming. It is better to have a one-shot article, than no articles at all!

You can write short articles on many topics: highlighting how you and your group use SAS, about what is new with SAS, about corporate SAS support, about new SAS products, about using SAS on different company computing platforms, how-to tips and hints, and so on. Use your imagination and try to come up with a fresh topic for newsletter articles, instead of recycling old ideas. There are so many facets to SAS that this should not be a hard thing to do. Make sure you are writing with a corporate slant so that your users can immediately put the information to use within your organization's SAS infrastructure. You want the information you present to be both educational and useful. That is how you can build a dedicated readership.

If you are a reliable source of material, you can count on being published often. If you have a lot of ideas, consider asking the editor if she would like to feature a "SAS Corner" page in the newsletter. You can populate that page with short articles on SAS tips and techniques. You can also publish announcements important to the SAS community such as the release of hot fixes, maintenance releases, and new SAS products that were recently purchased. As stated earlier, the announcements are the easy part. If you want to have a successful readership, you must make sure your article content is technically accurate. And, of course, make sure your name and contact number are visible on the page so that people link you with all things SAS in your organization.

Create an In-House SAS Listserv

E-mail-based listservs have long been popular vehicles for IT professionals to trade programming tips and techniques. The way a listserv works is that prospective members send the listserv owner an e-mail requesting participation in the listserv. The owner subscribes the members to the list. Then, when any list member sends an e-mail to the listserv, *all* subscribers receive a copy of that e-mail. Consequently, all members participate in the e-mail discussions held on the listserv.

Consider setting up an in-house SAS listserv. You could use commercial listserv software, have a web-based system, or use your organization's e-mail system. To do this, the obvious first step is to get permission from the relevant upper management for you to establish the in-house SAS listserv. Then, you must work with either someone in your systems group or in the e-mail administration group to get the listserv set up. When working with the technical staff, inform them that the goal is to have an in-house listserv. They should know the best way to set that up.

The way your in-house SAS listserv can work is to have users post their SAS questions and relevant documentation such as log snippets to the listserv. Experienced, knowledgeable SAS programmers such as you respond by sending specific answers, or tips and hints on how the poster might solve the particular problem. Such responses from well-informed SAS professionals usually provide enough information for the originator to get such issues resolved. If a particular problem cannot be easily solved or looks like a SAS error, you take the initiative and route it to the SAS liaison so that she can relay it to SAS Technical Support. When the SAS liaison replies with the solution from SAS Technical Support, post the solution to the listserv so that all users can benefit from it.

As the person who establishes the SAS help listserv, you need to check it often and make sure you are addressing your colleagues' questions in a timely manner. The only way the listserv will gain traction as a collaborative vehicle for trading information about SAS is if people are being responded to in a reasonable amount of time. If they are working on a project with a deadline and need an answer to move forward, they might not have the luxury of waiting a day or two. So, monitor the listserv and make sure your coworkers are getting the feedback they need on their SAS technical questions in an expeditious manner.

Your in-house SAS listserv is also a great vehicle for communicating corporate SAS initiatives to all SAS users in the organization. You can post SAS software announcements such as hot fixes, version upgrades, server replacements, SAS server reboots, and new SAS software that was recently purchased. You can post updates on the progress of resolving SAS issues that affect numerous staff members. The listserv is also a good place to post the SAS tip of the week that you write. Use your imagination and determine what SAS information would be of particular interest and use to your colleagues. Then, package it in an email and post it to the listserv.

An in-house SAS listserv provides a lot of advantages, not all of which are readily apparent. It obviously offers staff a convenient space in which to ask SAS questions and receive answers. The listserv also promotes a sense of unity and camaraderie among the SAS professionals in an organization. It provides a great forum for discussing the best ways of performing various tasks and the merits of using various SAS constructs to get work done. People learn a lot when reading different solutions proposed by various staff members to the same problem. Listserv postings make educational and entertaining reading for IT staff interested in widening their SAS programming horizons. Finally, the listserv provides you with a great forum in which to demonstrate your SAS expertise and reinforce that you are the top SAS programmer in the organization.

Start a Quarterly In-House SAS Newsletter

How hard would it be to start a quarterly in-house SAS newsletter? Not hard at all, really, because you are going to make it an e-mail-based newsletter. So, you need to obtain a list of the e-mail addresses of the SAS users in your organization. Then you must devise relevant and interesting SAS content for the newsletter. Doing so should not be as hard as you might imagine. If you are only "publishing" four times a year, you have a lot of time to plan the content of the newsletter. So, start planning your very first issue of the "XYZ Corporation SAS Newsletter" right now.

Here are some of the things you could include in the SAS newsletter:

- a brief article about a particular facet of SAS (for example, a feature of PROC DATASETS)
- tips from other SAS users in your organization
- what's going on in your organization and the greater SAS programming world
- links to interesting SAS users group papers
- information about recently purchased SAS products
- notices about hot fixes and maintenance releases
- announcements of in-house SAS users group meetings

There are a number of ways that you can obtain content for the newsletter. The best is to ask your SAS programming colleagues to contribute articles and tips. It is best because what they contribute is most likely to be relevant to other programmers in your organization. Another idea is to look at what other SAS users groups put in their newsletters. If you perform a web search, you will see that a number of local SAS users groups publish newsletters. Subscribe to them so that you can look through them for ideas on what might work for your own newsletter. Finally, comb through online SAS users group proceedings. You will find dozens of ideas on topics that would be of interest to your coworkers. Also, you might want to simply write up a quick summary of a particular paper in the proceedings with a link to that paper so that staff can read it. Anything that is informative to SAS programmers is fair game for your newsletter.

Make certain you lay out the newsletter so that it is readable with lots of white space between paragraphs and double-check the grammar and spelling. Each issue should be as interesting and informative as you can make it. Keep in mind that the goal is to provide relevant SAS information the "average programmer" might not necessarily know or be aware of. You are also providing a strong voice for promoting SAS programming in your company. By doing so, you help build a sense of camaraderie and esprit de corps among the SAS professionals in your organization. They will end up looking forward to the next issue of your SAS newsletter.

Send Out the SAS Tip of the Week

A great way to highlight your SAS expertise while providing a service for all of your colleagues who use SAS is to publish a *SAS Tip of the Week.* The SAS Tip of the Week is an e-mail that provides a concise tip on using a particular facet of the SAS programming language. It doesn't have to be something profound--just something your coworkers will find useful and probably would not have thought of. The tip itself should be one-to-two paragraphs of explanation followed by an illustrative SAS code example. The code example should be crafted so that it can be cut and pasted into SAS Display Manager or SAS Enterprise Guide and run. This enables your constituents to try it out for themselves to better understand your point.

Considering the breadth and scope of SAS, you should have lots of ideas for SAS tips. Mine your SAS programs for the clever ways you are using SAS statements, procedures, functions, macros, data set options, system options, formats, informats, and so on. There is sure to be a lot of material there. Next, consider combing through SAS documentation for ideas. The DATA step references in the online SAS documentation found on support.sas.com are especially great sources to mine for ideas. SAS users group proceedings provide another hot source for finding SAS tips and techniques. So, browse through them for relevant ideas. Finally, ask your coworkers if they have an idea or two for the SAS Tip of the Week. You undoubtedly work with a lot of smart people, so they should have something useful to share.

Your SAS tips must have integrity because they are going out to SAS users of varying abilities. So, you need to ensure that you fully understand and explain the concepts in your tip. Read and reread your text to verify that it concisely explains the tip. You should lay out the issue, describe the relevant SAS code that resolves the issue, show the SAS code, and then take the reader through the code so that the reader understands how it resolves the issue. Keep your explanations short and to the point; it is a SAS tip, not a SAS novel. Where appropriate, include a link to where the reader can find more information about the particular SAS programming language construct illustrated in the tip.

Here is an example of a tip:

All,

Did you know that you can improve the look of your reports that are generated using SAS by adding a little color to the titles?

You can use the Output Delivery System (ODS) ESCAPECHAR statement and inline formatting to add and change colors within your report titles. Here is an example:

```
ods escapechar = "^";

ods pdf file="example2.pdf";

title1 "^S={color=blue}Though this title started off blue ^S={color=red}it
changed to red";
title2 "^S={color=green}This title was green ^S={}before changing to black";

proc print data=sashelp.class;
run;

ods _all_ close;
```

We start off with the ODS ESCAPECHAR statement, which defines a special character that will be used to signal to SAS that we want to perform inline formatting. In our case, whenever SAS sees "^" in a title statement, it knows that ODS inline statements follow. Thereafter, we put "^" to work in two TITLE statements, each time specifying the color for the following text. If you cut and paste the example above into a SAS Display Manager session and execute it, you will note the changes in the **colors** in the title lines. Really, makes you think about how you may be able to spruce up those tired old reports, doesn't it?!?!?

The **SAS Technical Paper of the Week** provides a lot more information on this particular topic:

Let the ODS PRINTER Statement Take Your Output into the Twenty-First Century
Scott Huntley, SAS Institute Inc., Cary, NC

http://www2.sas.com/proceedings/sugi31/227-31.pdf

Go ahead; give those color printers a run for their money and improve the look of your project's reports!

Best of luck in all your SAS endeavors!

Sincerely,

Michael A. Raithel

Pick a particular day of the week to send your SAS Tip of the Week and send it out to all corporate SAS users every week on that day. Once you start, do not change days or miss a week. You want to build up a following, and faltering will hamper that endeavor. A good trick to keep ahead of the game is to sit down and write enough tips for four or five weeks in a single sitting. Thereafter, try to have at least five weeks of drafted SAS tips ready to go. This will tide you over when project emergencies eat up all of your time, and ensure that the SAS Tip of the Week goes out on time.

Volunteer to Extract Data

One way to increase your SAS reach is to volunteer to extract data from corporate data sources for groups that don't use SAS. Your organization probably has data landlocked in many different file types and in relational databases. Corporate operational data can often be found in fixed-length or variable-length flat files, tab-delimited files, XML files, relational databases, and HTML files, to name a few. Many of these data stores lie behind production applications and contain valuable business information. Most business analysts in your organization probably don't have the programming skills and programming tools necessary to access these types of files, extract the data, and load it into the spreadsheet software they are familiar with. Consequently, they are missing out on opportunities to analyze the data.

Fortunately for you, SAS excels at accessing data from various and sundry types of data sources. Base SAS enables you to process many types of flat files, including pipe-delimited files, tab-delimited files, HTML files, TCP/IP sockets, and XML files. The SAS/ACCESS products facilitate processing a wide variety of data sources. SAS/ACCESS Interface to PC Files provides you with the ability to access data in Microsoft Excel files and Access databases. Other SAS/ACCESS products offer facilities for accessing tables in specific relational databases such as Sybase, SQL Server, and Oracle. So, having Base SAS and a few of the SAS/ACCESS products puts you in a great position to do a lot of good for your organization.

Be proactive and look around your organization for opportunities to use SAS to extract data from the back-end files and databases of major applications. Seek out business analysts and ask them if there is specific information that they would like to get from various applications. Work with them and with the IT staff who designed the systems to determine the type, structure, and content of the back-end data repositories. Then, craft SAS programs to extract the data, format it, and output it to the file types the business analysts are the most familiar with. Many of your efforts are likely to be one-shot data extracts for the analysts. However, look for opportunities to make your programs production-friendly so that they are scheduled to run on a cyclic basis in cases where the analysts want continued periodic extracts of data that are frequently updated.

It might not always be your organization's business analysts who could use some help. Sometimes systems staff do not have the proper programming tools to manipulate back-end data to the point where they can analyze it. Or, they might have the tools, but it might be tedious because those tools are not as powerful as SAS. For example, the systems staff in one organization needed to analyze the types of spam mail that were trapped by their spam filters. The information was stored in very, very large log text files. The systems staff did not have an easy way to process those voluminous files to find the specific keywords they were looking for and count the instances in which they occurred. A SAS programmer got involved. The systems staff supplied the programmer with a list of keywords, the record layout, access to the spam filter log files, and the desired information that was needed. The programmer used Base SAS to process the log files and produced the information needed for the spam log analysis. This effort saved the systems staff a lot of work and the SAS program is still in use.

Opportunities for you to extract data from many different systems in your organization are out there. So, extend your SAS reach by proactively looking for them. You can do a lot of good for your organization while making a name for yourself as the top SAS programmer.

Set Up a Data Quality Regimen

Many organizations receive data from outside parties such as customers, clients, grantees, business partners, vendors, and other organizations. Such data is usually transferred in files with an agreed-upon file structure and file type. They could be Excel files, XML files, CSV files, pipe-delimited flat files, or a host of other file types. Whatever the file type, the data in the various fields are expected to be in a specific order, to have a specific data type, and to have values within specific ranges. Although submitting organizations always try their best to make sure their data follows the agreed-upon protocol, there can be factors that cause their files to be substandard. Some common reasons include misunderstanding the file layout, producing systems errors when creating the file, including incomplete data in the systems the file was extracted from, and duplicating records due to multiple attempts to build the submission file. Because of this, it is incumbent upon the receiving organization to validate the data files they receive from out-of-house before considering processing the data stored in them.

Fortunately for you, SAS is the perfect tool for evaluating new data sets. Therefore, consider setting up a data quality control regimen for in-house projects that regularly receive data from outside sources. To do this, you need to write a SAS program that analyzes the specific type of data set your organization is expecting. Then, when new data sets come in-house, you run the program against the data set and provide the results to your users so that they can either accept or reject the data sets.

To get started, the business analyst responsible for the data files needs to provide you with the record layout. Next, the analyst should specify the various checks that must be done to validate that the data fields meet the requirements. The analyst should give you specifications on the type of report to be produced; for example, should you list every data anomaly or have a summarization of the discrepancies? Finally, you should be given security permissions to access the directory the files are stored in.

Sometimes the business analyst might not have a clear idea of the issues one should check for in an incoming file. So, you might have to use your own best judgment and suggest some data anomalies to be wary of. Here are some to consider:

- duplicate records
- missing records
- invalid record types
- too many data fields
- missing data fields
- data fields not long enough

- data fields that are too long
- data fields that are not in the correct order
- duplicate primary key variables (for example, two records with the same Social Security number)
- invalid data in primary key variables
- primary key value is not found in the database
- character data in numeric fields
- numbers out of range (for example, ages greater than 150 years)
- missing values

This list is hardly exhaustive because there are countless ways that a data file can get messed up. However, it provides good starting points for suggestions that you can provide to the business analyst you are working with. Once the two of you begin talking, other data integrity rules will doubtless come to mind. A great source of information about data cleaning issues is SAS Press author Ron Cody's book, *Cody's Data Cleaning Techniques Using SAS, Second Edition.* If you are going to be doing a lot of work with files that might have dirty data in them, this idea-packed book is a must.

When you have the specifications, you can write your SAS program. It is a good idea to modularize your program so that each specific data check is separate and well-commented. Doing this will help you to more easily modify your program when requirements inevitably change, causing specific edits to be altered or removed. When you get to the section where you are going to write the report, keep your audience in mind. Make sure to create a very professional-looking report using the Output Delivery System. Pay attention to headers and footers to make sure they are accurate and meaningful, and that there are no spelling or wording errors. Put the filename and the date received into the headers, and have your contact information in the footnote. It is a good idea to create the report as a PDF, especially if it is going out-of-house, so that it cannot be accidentally modified.

Once your program is written, tested, and has successfully analyzed the first data files, start planning on how you can put it into production. Work out a routine with the business analyst so that the two of you work as a team to vet incoming data. The routine might look like this:

- The business analyst gets a file from the client and lets the client know the file was successfully received.
- The business analyst sends you an e-mail with the name of the file and its location.
- You run your QC program against the file and create the QC report.
- You send the QC report to the business analyst.
- The business analyst sends the report to the client specifying whether the file had issues.
- If the file had issues, the client produces a new file and sends it to the business analyst, and the cycle repeats itself for the new file.

You can deviate from this routine if you are able to become the person who receives the data files from the client. In that case, you would perform the functions specified for the business analyst, as stated above. However, you would also take the step of moving fully qualified data files into a directory that the business analyst had access to. Or, you could write a SAS program that processes the incoming files and turns it into a format the business analyst can use, such as an Excel file or an Access database. There are many possibilities, but it is best if you try to take on as much of the work as possible, using your full bag of SAS tricks.

Be a Team Leader or a Project Leader

It is no secret that in most organizations, assuming more responsibility is the way to increase your visibility and your paycheck. For programmers this means becoming a team leader, or a project leader, or a manager, depending on the specifics of a particular organization and the opportunities at hand. People in leadership positions often perform more organizing and planning activities and less hands-on programming. They tend to spend more of their time in project planning tasks and in meetings than programmers do. So, hard-core SAS programmers often wonder if promotions to management positions will dampen their programming skills and make them less technically proficient as time passes. Sometimes there is tension between wanting to accept more responsibility, but not wanting to lose your technical edge.

If you make SAS programming both your hobby and your profession, then you do not have to worry about promotions dampening your SAS skills. As long as you stay connected to what is new with SAS software and programming techniques, you will continue to be proficient at SAS programming. When you take on management responsibilities—which you should do as a natural part of your professional progression—make sure you still have a hand in doing some programming for a particular project. That will keep your SAS programming skills sharp.

There are other benefits to becoming a team leader or a project leader besides better visibility and additional pay. Leadership positions enable you to interact more with corporate management and with clients. This gives you opportunities to understand their needs and to develop your own management style. It also provides you with a better overview of the work at hand and what is happening in your organization. It is easier to understand the big picture when working with upper management than it is when working down in the trenches as a programmer.

Leadership positions give you the opportunity to get things done the way that you want them to be done. Sooner or later, you are likely to have been frustrated by the way some manager has run a particular team or project. When you become a leader, you can organize the work the way that you think is most expedient. You can put your own managerial stamp on the projects you manage, so they are done to your own liking while matching the requirements.

Becoming a leader enables you to follow the progress of other SAS programmers and to mentor them. You can provide guidance to their SAS training and expose them to information and to classes that make them stronger SAS programmers. That is good for them, good for you, and good for your organization.

Many companies provide classes on how to manage staff, so take advantage of the management training your organization has to offer. Do not simply assume that because you are the top SAS programmer you are already a top manager. There is a lot to learn about managing other programmers and very little of it involves being technically proficient. Hence, you need to make sure you attend any corporate classes available to you. If your organization does not offer such training, see if you can take a course at a local college or community college. Or, perhaps there is an online management training class that your company will pay for. If none of these possibilities pan out, there is a plethora of books available on managing technical staff.

Working your way from programming to management is a natural progression in the career of an overachiever like yourself, so embrace the changes if that is what you want. You can always choose to continue to be a SAS programmer, letting promotional opportunities pass you by. Whether you stay a SAS programmer or become a leader is a question that only you can answer. Just know that with a little effort and perseverance on your part, you can indeed stay technical. Just because you get to boss around a group of SAS programmers doesn't mean you have to stop being the top SAS programmer in your organization!

Chapter 7: SAS Users Groups

Introduction

Imagine yourself walking down the hallway in the ballroom area of a swank hotel. You are dressed in business casual attire and have a badge with your name and the name of your organization hanging from a lanyard around your neck. You have a newly issued canvas tote bag that contains a booklet with the abstracts and the scheduled times of dozens of SAS presentations slung over your shoulder. You have just attended a 50-minute presentation on a particularly interesting SAS programming topic and are on your way to another lecture. Or, better yet, you are on your way to give your own 50-minute presentation on a cutting-edge SAS programming technique.

The hallway around you is filled with SAS programming professionals of all types. There are programmers, programmer-analysts, systems analysts, statisticians, researchers, business analysts, managers of SAS programmers, and SAS staff strolling to presentations or simply milling about.

As you walk, you hear snatches of conversations about dozens of SAS topics: DATA steps and PROC steps; hash objects; statistical procedures; SAS releases, versions, and hot fixes; SAS macro programming; the Output Delivery System; moving files between operating systems; reading various file types into SAS; SAS formats and informats; SAS functions and CALL routines; SAS data set options and SAS system options; SAS solutions and SAS Analytics; and the correct syntax of various SAS statements. People are conversing openly without once feeling a bit silly talking about saving a few seconds of CPU time or a few thousand I/Os, or whether it is better to use the NODUPKEY or NODUPREC option for a particular sort, or of the best way of using the Output Delivery System to create publication-ready output. Everybody is perfectly at home discussing such SAS topics.

Does all of this sound farfetched? Well, it isn't. Every year, dozens of SAS users groups hold half-day, one-day, two-day, and even three-day conferences. The convention venues range from conference rooms in local businesses, to meeting rooms at colleges and universities, to ballrooms and meeting spaces in major hotels. Depending on the size of the SAS users group, the number of attendees can range from a few dozen, to several hundred, to several thousand. Attendees work in contract research organizations, banking, finance, the federal government, local governments, pharmaceuticals, colleges and universities, utilities, insurance companies, for SAS Institute, and for a host of other types of businesses. At smaller conferences, participants usually come from the local area; at larger ones they come from the surrounding region; and at still larger conferences they come from all over the country and from around the world. Conference attendees are all a part of the larger community of SAS programming professionals, drawn together to share the latest information about SAS programming and to learn from one another. And you should definitely be among them.

Attending a SAS users group meeting or conference can be a very rewarding experience far beyond just what you learn about SAS. Think how much fun it would be to speak SAS and have everybody within earshot understand exactly what you are talking about. While you are at a SAS users group conference, you feel like you are part of a profession that is bigger than the job that you do. That is a feeling that you cannot easily get at work—that you are part of something big; that you are part of the international SAS community. Because of that commonality, you can make lifelong friends and professional acquaintances at SAS users group conferences. You can also make a name for yourself as one of the top SAS programmers. So, attending SAS users group meetings is a must.

SAS Institute Inc. recognizes four types of SAS users groups: in-house, local, regional, and special interest. In addition, there is an international users group named SAS Global Forum. All SAS users groups are run by volunteers from the SAS programming community. SAS provides support to users groups, consistent with the size of the groups and the frequency of their meetings. For example, SAS might provide speakers from the company, SAS Press books as giveaways, e-mails and marketing promotions for the meetings, and giveaway items such as SAS coffee mugs, USB ports, or pens for smaller users groups. They might provide the aforementioned support as well as staff to operate a demonstration room for larger users group meetings. However, the meetings are organized and run by SAS users for SAS users; SAS is an invited guest.

This chapter discusses the various types of SAS users groups available to you. It provides an overview of each type of SAS users group: in-house, local, regional, special interest, and SAS Global Forum. These sections should give you a good idea of the composition and routine of each type of meeting. While reading this chapter, think about which types of SAS users groups you would like to participate in and how you can make your own contribution to them.

In-House SAS Users Groups

Overview

Organizations with an embedded community of SAS programming professionals often form their own in-house SAS users groups. The organizers identify a list of potential attendees, choose a name for the group (for example, The Zephyr Corp SAS Users Group), and register it with SAS via a registration form found on support.sas.com. Once the in-house group has been registered, organizers arrange meetings on a schedule that is convenient for them and their attendees. The organizers reserve the meeting rooms, choose the speakers, and publicize the events to their colleagues who use SAS. SAS Institute supports such meetings by annually supplying them with SAS speakers, SAS Press books, and other SAS giveaways. There are currently over 140 in-house SAS users groups registered in the United States.

Organizations are free to run the in-house users group meetings when they want and for the duration that best suits them. However, the most common pattern is to hold a half-day meeting each quarter. This tends to have a minimum impact on the participants' time away from work, and allows enough time for the organizers to find speakers. Speakers commonly come from SAS programmers within the organization. However, some in-house SAS users groups might occasionally invite well-known SAS presenters from neighboring organizations. Whatever the case, the meetings commonly have the following agenda:

- Introductory remarks from the organization's chair or steering committee.
- Announcements about SAS versions and technical support from the in-house SAS support staff.
- Announcements of upcoming SAS training.
- The SAS technical presentations. These are largely PowerPoint presentations, but might include switching to SAS Display Manager or SAS Enterprise Guide to show code executing in real time. Presentations are generally an hour long and are followed by question and answer sessions about the material that was presented.
- A general question and answer session where people ask SAS- programming questions, and other attendees provide suggestions, tips, or answers
- A reminder of the date of the next meeting and who will be presenting.

In-house SAS users groups provide a convenient forum for SAS professionals to learn from and network with others within their own organization. Presentations from colleagues are likely to be related not only to the business the organization is engaged in, but also to the specific way that the organization actually pursues that business. So, topics such as using SAS on the company's various computing platforms, extracting data from corporate databases, and discussing issues related to organizational data security might be an integral part of the presentations. Consequently, attendees can learn about SAS programming techniques as they relate to the data, the computing infrastructure, and the business of their own organization.

How You Can Participate

If your organization does not already have an in-house SAS users group, then start one. Information about the best way to approach this task can be found in the section **Start an In-House SAS Users Group** in Chapter 6, "What You Can Do in Your Own Organization." Starting an in-house users group puts you in a leadership role and enables you to do something beneficial for the SAS professionals in your organization. Can you say "win-win situation"?

If an in-house SAS users group is already established, make sure you are a regular attendee at the meetings. Learn about your colleagues' SAS practices as they relate to your business and to your IT infrastructure. It will make you a more effective SAS programmer within your company. But, don't just listen; contribute. Come up with a relevant SAS presentation and tell the organizer that you want to be put on the schedule. Since you are (or will soon be) the top SAS programmer, make sure you are a frequent presenter at the meetings. You should be giving presentations and taking an active role in question and answer sessions. If you are your site's SAS Administrator, arrange to have about five minutes per meeting to say something about the latest releases of SAS and the plans for future migrations to newer releases. If you triage SAS technical issues for your establishment, see if you can have 10 minutes to discuss a common or an interesting SAS support issue and its resolution or workaround. Volunteer to run any meetings where the organizer will be out of town. See if you can help the organizer by recruiting fellow SAS programmers to present, or by recruiting guest speakers from other nearby organizations. Overall, be an active, vocal, recognized contributor to the meetings.

Local SAS Users Groups

Overview

Local SAS users groups provide opportunities for SAS users within a specific geographic region to meet and learn new things about SAS programming. The geographic area can range from a city, to a county, to areas within a state, or to an entire state, depending on the area's population density. Local SAS users groups are started when a band of SAS professionals from one or more organizations, establish an executive committee, choose a name for the group, and then work with SAS to officially register the group. SAS supports a local SAS users group by annually supplying it with SAS speakers, a shipment of SAS Press books, and other SAS giveaways. Membership is open to all SAS users who fall within the given geographic area the users group serves. There are over 70 local SAS users groups in the United States.

Most local SAS users groups hold quarterly meetings that are either half-day or full-day events. Meetings are usually held in conference rooms of member organizations, but are sometimes held in hotel meeting rooms if the group is large enough. Many local groups require members to officially register and some charge a yearly registration fee. The registration fee covers expenses such as meeting promotion, paying for speakers' travel expenses, renting a hotel meeting room, and funding the continental breakfasts and snacks commonly served at the meetings. Once members are registered, they receive e-mail notifications of upcoming meetings.

A typical local SAS users group meeting might follow a schedule like this:

- Continental breakfast and time to meet and greet fellow attendees.
- A call to order and introductory remarks from the chair or members of the steering committee.
- Announcements of upcoming meetings and speakers.
- Remarks from the treasurer concerning how to officially sign up for the group and when membership fees are due.
- A SAS technical presentation, followed by a question and answer session about the presentation.
- Coffee break.
- Additional SAS technical presentations.
- A general SAS question and answer session where participants ask the group general "how-to" questions and members of the audience respond with their own insights.
- A reminder of the date of the next meeting and who will be presenting.

At full-day events, attendees usually have lunch at nearby establishments or in the host organization's cafeteria. Sometimes there is a specific theme for the presentations held in the morning and those held in the afternoon. For example, morning presentations might involve various techniques for efficiently processing large SAS data sets, while the afternoon presentations might focus on SAS statistical programming topics. Participants are free to attend as many or as few presentations as they want, depending on their personal interests and work schedules. Some local one-day conferences have grown so large that they also have concurrent sessions. So, attendees can choose the topic they would prefer to hear about from two or more concurrent sessions.

Local SAS users group meetings bring together a disparate group of SAS programming professionals with varying skill and experience levels from diverse area organizations. They can be great places to learn new things about SAS programming and to find out what other organizations in the area are doing with SAS. They are also good venues for networking with SAS programming professionals in your local area. Top SAS programmers should be actively involved in local SAS users groups.

How You Can Participate

The easiest way to participate in a local SAS users group is to simply show up and attend the meetings. You can determine which local users groups are in your area by accessing the users groups page on the SAS website http://support.sas.com/usergroups/. On that page, click on **U.S.** in the navigation tree on the left to go to the SAS Users Groups in the US page. See Figure 7.1[1].

Figure 7.1: SAS Users Groups Page

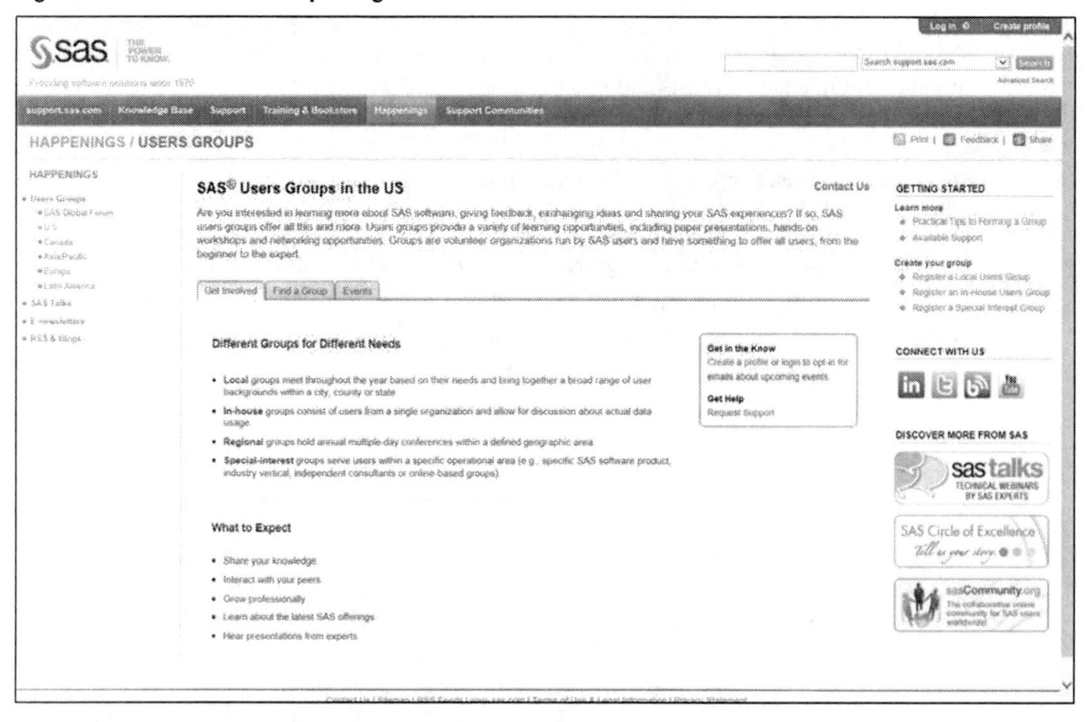

Click the **Find a Group** tab to see a map that breaks the country into regions. See Figure 7.2.

Figure 7.2 Regional SAS Users Groups

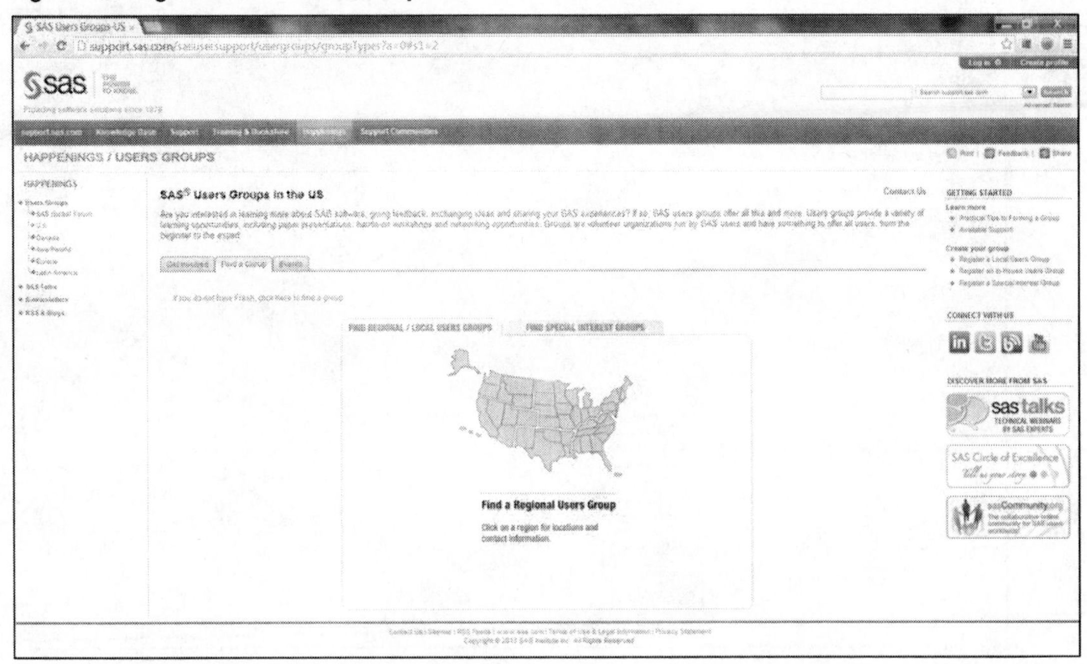

When you click on a region, you get a drop-down list of the states within that region. When you click on a state, contact information for the local SAS users groups within that state is revealed. (See Figure 7.3.) Send an e-mail to the contact expressing your interest in joining. Ask to be put on the group's e-mail list and, if available, the snail mail list, too. Be sure to inquire about the URL for the group's website and whether there are annual dues. (Don't forget that many organizations—hopefully your own—reimburse employees for membership in professional organizations associated with their work.)

Figure 7.3 Northeast SAS Users Groups

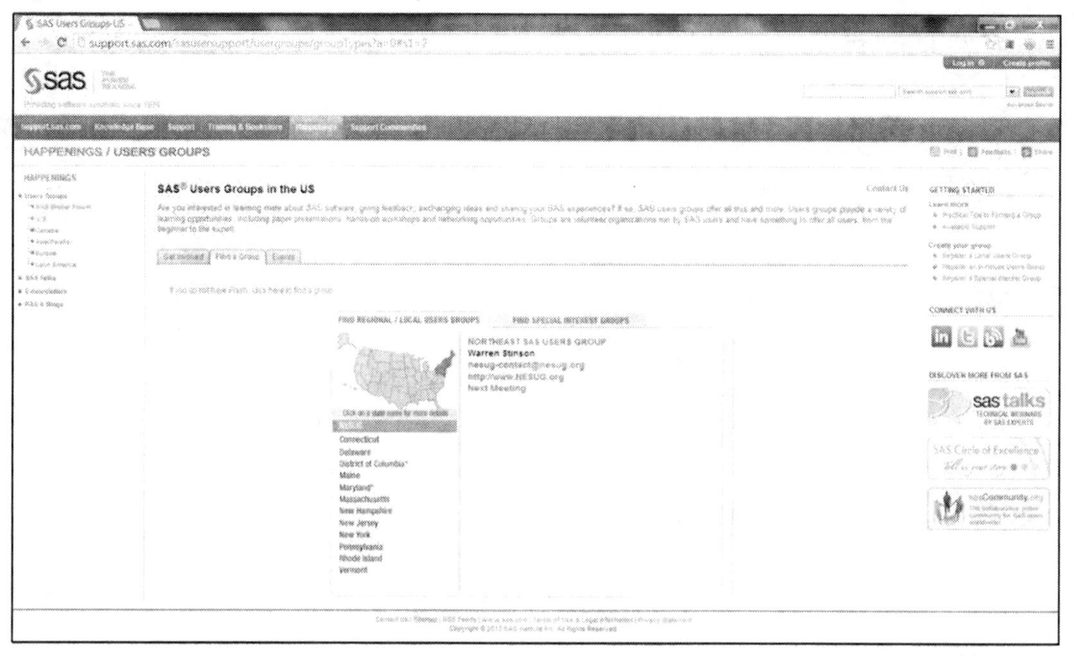

The next logical step for participation in local SAS users groups is to give presentations. Local users groups are always looking for good presentations to fill their meeting calendars. With your top-notch knowledge of SAS, you shouldn't have much trouble coming up with a good idea for a presentation that appeals to a large group of SAS programmers. You might consider giving a hard-core SAS technical presentation during which you demonstrate capabilities that many people might not be acquainted with. Or, you might present a tutorial on a step-by-step solution to a programming problem using SAS. Another idea is to capitalize on the main industry of the area (for example, insurance in the Hartford, Connecticut area, or pharmaceuticals in the Philadelphia area) by showing how you can approach a common industry-related programming problem with SAS. Whatever the case, be creative and come up with an abstract and an outline of the presentation. Then, approach the local SAS users group organizers with your idea. A little initiative on your part will doubtlessly yield benefits!

SAS users groups are run by SAS professionals like you, so consider volunteering to help run a local SAS users group. Local groups often have appointed or elected "officers" such as chairperson, secretary, SAS liaison, treasurer, sponsor liaison, and registrar. So, see if you can fill one of those positions and lend your considerable talents to the organization. It will involve some of your free time, but you will be making a great contribution to the other SAS professionals in your area.

Another way to help the group is to provide space for one of their meetings. Many local users groups find it challenging to line up venues in which to meet. Meeting spaces must be big enough to accommodate the group, local enough so that transportation to them is not a hassle, and readily accessible to visitors from outside the hosting organization. If your employer has a meeting space large enough to accommodate your local users group, ask management if they would be amenable to hosting at least one meeting per year. If so, contact the users group organizers and work with them to get an upcoming meeting scheduled at your site. Make sure the organizers publicly recognize you and your organization's largesse in the announcements at the beginning of the meetings.

Regional SAS Users Groups

Overview

Regional SAS users groups serve SAS professionals across multiple states grouped into logical geographic regions. SAS users who fall within a given geographic area automatically "belong" to that particular regional SAS users group. There are five regional SAS users groups in the United States:

- The NorthEast SAS Users Group (NESUG) – www.nesug.org
  - Serving Connecticut, Delaware, District of Columbia, Maine, Maryland, Massachusetts, New Hampshire, New Jersey, New York, Pennsylvania, Rhode Island, and Vermont
- The SouthEast SAS Users Group (SESUG) – www.sesug.org
  - Serving Alabama, District of Columbia, Florida, Georgia, Kentucky, Maryland, Mississippi, North Carolina, Puerto Rico, South Carolina, Tennessee, Virginia, and West Virginia
- The Midwest SAS Users Group (MWSUG) – www.mwsug.org
  - Serving Illinois, Indiana, Iowa, Kansas, Michigan, Minnesota, Missouri, Nebraska, North Dakota, Ohio, South Dakota, and Wisconsin
- The South Central SAS Users Group (SCSUG) – www.scsug.org
  - Serving Arkansas, Louisiana, New Mexico, Oklahoma, and Texas
- The Western Users of SAS Software (WUSS) – www.wuss.org
  - Serving Alaska, Arizona, California, Colorado, Hawaii, Idaho, Montana, Nevada, Oregon, Utah, Washington, and Wyoming

All regional SAS users groups are non-profit, educational organizations that host annual SAS conferences that take place over two or three days in meeting facilities at major hotels. Think of it as a big sleepover for SAS programmers. Regional conferences are big events that attract significant numbers of SAS professionals. There might be from 250 to more than 800 SAS users, depending on the location and the number of SAS users in the specific region. People do not have to reside within the particular region to attend that region's conference—they are open to all SAS users. Often enough, some presenters and attendees come from outside the region to attend a regional conference. And, some of the regionals even draw international SAS professionals.

SAS users residing in the region who are on the regional users group's mailing list, or on SAS Institute's users groups mailing list, normally receive e-mail notifications of the upcoming conference. The e-mails provide general information about the conference and reference the regional users group's website where more detailed information can be obtained. SAS users access the conference information pages on the website and complete the online registration if they decide to attend. Attendees pay a registration fee of several hundred dollars and are offered a discounted fee to stay at the conference hotel. Various half-day and full-day SAS classes taught by either SAS staff or by others with SAS expertise are often offered as extra-fee items the day before or the day after the conference. SAS certification exams are also sometimes held just before or after the conferences. Conference goers typically register for such classes and certification exams online when they register for the conference.

Regional conferences are organized into *sections* based on subject matter. Concurrent sessions of the sections are held throughout the conference. This means that at any given time during the day, there are several technical presentations going on. So, attendees have a lot of choices and can attend presentations they believe will best further their own SAS knowledge. For example, a conference might have the following sections:

- Analytics and Statistics
- Beyond the Basics
- Business Intelligence
- Coders' Corner
- Data Presentation and Reporting
- Hands-on Workshops
- Tutorials

A detailed schedule of the dates and times of the various presentations is normally posted on the regional SAS users group websites. They also provide abstracts[2] for the papers that will be presented. Attendees can access the schedule, read the abstracts, and customize their itinerary before the conference. Some conferences also offer mobile apps for smart phones and tablets. You can create a customized schedule using the mobile app prior to attending the conference. Alternatively, participants can wait until they arrive at the conference and are given a packet of materials at the registration desk. That packet contains a schedule and the abstracts for the papers that will be presented at the conference. It is common to see conference goers walking around with schedules that have bright yellow streaks on them from where various presentations were highlighted for attendance.

Conference presenters also write a technical paper on their topic that is published in the *conference proceedings*. The conference proceedings are a compendium of all of the papers presented at the conference, arranged by section. The regional users groups publish the conference proceedings on their websites so that all SAS professionals can freely access them over the Internet. Because many papers are written by SAS professionals from the user community and by SAS staff, the proceedings are a treasure trove of SAS programming tips, techniques, and how-to information[3].

Another key feature of SAS users group conferences is the SAS Demo Room. This is a large meeting room that contains workstations set up by SAS staff to demonstrate various SAS products and capabilities. SAS personnel are on hand to answer questions about any facet of the software. If possible, they will demonstrate the answer to a question or the methodology for doing something with SAS firsthand. If not, they will write down the question and the questioner's contact information so that they can research the problem and respond back with the answer after the conference. The SAS staff also holds live demonstrations at various scheduled times throughout the day in the demo room. Demo rooms always contain a SAS Press area where conference goers can purchase SAS books at a discounted price and often talk to the authors about their books. The SAS Demo Room provides a convenient place for hanging out between interesting presentations and learning more about the types, uses, and intricacies of SAS software.

SAS regional users group conferences also frequently include the following features:

- **Opening Session** The opening session is usually the first big group meeting of a conference and is often held the night before the conference. The conference chair formally welcomes everybody to the conference, introduces the section chairs, provides an overview of the conference, and publicly recognizes the generous corporate sponsors. A keynote speaker gives a presentation on a SAS or business topic. Then, with final well-wishes from the chair, attendees normally head out to a get-acquainted mixer or opening night mixer.
- **Code Clinic or Code Doctor** SAS programmers are on hand to answer real-life SAS programming questions. Attendees are encouraged to bring in problem programs and questions so that the experts can review them and proffer solutions to the problems. Code clinics are often manned by some of the top SAS programmers in the region or in the country, so you are sure to get very solid answers to your questions.
- **Hands-on Workshops** These tutorial-style workshops enable attendees to sit down at workstations and learn new SAS programming tools and practices first-hand. They frequently present cutting-edge SAS programming techniques and ideas and usually run about ninety-minutes long. Access to hands-on workshops is included at no additional charge, and they are *very* popular. Conference goers usually line up for them early to make sure they get in. Hands-on workshops are a great way to take SAS software you don't already license in your organization for a test drive.
- **Kickback Party and Volunteers' Party** These types of gatherings allow attendees to meet one another, meet conference organizers and paper presenters, and meet SAS staff. The social events offer great venues for trading SAS programming tips and techniques with the best and the brightest SAS professionals in the region. They are good events for networking and socializing.
- **Meetups** Meetups are after-hour meetings on special topics of interest to attendees such as a SAS Enterprise Guide meetup. There is usually a bulletin board near the conference registration desk with slips of paper on it with the names of small conference rooms and time slots for when they are available. People write down a topic and sign their name. Others sign their name too, meaning that they plan to attend. Then, the interested parties show up and conduct their own meeting as they see fit.

- **Posters** Regionals have an area set aside for large posters depicting a particular SAS technique. Posters are created by presenters who have contributed an idea to the posters section chair and had that idea accepted. They write a paper, and also distill their idea onto large posters that are then fastened to displays in a hallway or conference room. Poster presenters are available at certain times throughout the conference to greet attendees and describe their work as represented on the poster.
- **Closing Session** The closing session is the last event at a regional conference. At that session, the organizers thank the volunteers (section chairs, presenters, session coordinators, and so on) for all of their hard work and formally pass the reins to the next year's conference chairs. If contributed papers and posters have been judged, then they announce the "best paper" and "best poster" awards for the conference. There are usually raffles and many SAS giveaways, such as SAS Press books, at the closing sessions.

How You Can Participate

The most basic way for you to participate is to simply attend a regional SAS users group conference. Regionals are normally held in the fall, so be on the lookout for e-mails advertising the Call for Papers is open or registration is open. You can also access their websites to determine the exact conference and registration dates. Regionals typically offer a discount for early registration, so make sure you register early enough to take advantage of it. Most regionals also offer "team discounts" for when a sizable group from the same organization plans to attend (for example, five or more people). Collaborate with your colleagues to see whether there are enough of you going to the conference to qualify for a team discount; it could save your organization some money.

Some regionals also offer "academic scholarships" for qualified students and "junior professional scholarships" for professionals new to using SAS who meet certain criteria. The scholarships might include waived registration, workshops, travel and transportation, and other perks. Each regional conference administers its own programs, so benefits vary by conference. The scholarship benefits are usually clearly spelled out on the conference websites. So, if you fall into either of these categories, it is worth looking into. If you do not, determine whether you have a colleague who does and alert that person to the possibilities of applying for such an admission to the conference.

When you attend a regional SAS users group conference, be sure to participate in as many activities as you can. Study the conference schedule ahead of time and build an agenda of must-attend technical presentations designed to further your knowledge of SAS. Plan on attending all of the social functions to network with your regional SAS programming peers so that you can trade SAS tips and techniques, and IT business practices. If a pre- or post-conference workshop is in your organization's budget, take one and learn something new that you can use when you return to work. Spend time in the SAS demo room meeting the hard-working SAS technical staff and discovering the intricacies of new SAS software that might be of future value to your business. Browse the SAS Press books and talk to the authors at the conference to learn more about SAS. In short; talk SAS, SAS, and more SAS with everybody you can possibly meet at the conference. You should leave the conference recharged and full of enough new SAS knowledge to bring fresh techniques and value to your organization.

A second way to participate in a regional conference is to write a SAS technical paper and give a presentation. People who write papers published in the proceedings are also on the hook for giving a live presentation of the paper at the conference. Most presentations are done using PowerPoint, but sometimes examples are given using SAS Display Manager or SAS Enterprise Guide. There are five basic types of papers given at regionals:

- **Invited Papers** The section chairs invite leading SAS professionals to give a 50-minute presentation on a SAS topic. Topics can range from SAS Enterprise Guide, to a SAS/STAT topic, to a tutorial on PROC DATASETS. Invited presenters write a technical paper that can be up to 20 pages in length and is published in the conference proceedings. If you have a good idea for an invited paper, contact the section chair via the e-mail link on the conference website and pitch it. That is how many successful presenters start off—with a simple query that leads to that first great SAS technical paper.
- **Contributed Papers** Contributed papers are 20-minute[4] presentations that come from attendees who send their ideas for a technical paper to the conference. Conferences send out a "Call for Papers" via e-mail and a posting on their websites. Interested SAS professionals submit an abstract and an outline to the chair of the section they are interested in participating in (for example, Statistics, Beyond the Basics, and Coders' Corner). The various section chairs vet the submissions and select the top paper ideas, often conferring with the Academic Chair of their conference. They inform the successful submitters that their paper has been selected, and encourage the others to try again next year. Authoring a contributed paper is a good way to begin your SAS presentations "career"[5]. You definitely know something about SAS, can write down your ideas in a cogent, organized way, and can articulate them in a 20-minute presentation. So look through your regional's previous year's proceedings and see which topics are covered and which are not. Can you present something interesting that has not already been covered? Write an abstract and an outline, and submit them to the conference before the paper submission deadline. Then, get ready to write your paper and create your PowerPoint presentation when you get the good news that your idea was accepted.
- **Coders' Corner** These are usually 10-minute talks that highlight a specific SAS tip, trick, or methodology. The conference call for papers also solicits coders' corner papers. This is another area where you can easily showcase your SAS programming abilities. Look through your SAS programs for a killer technique that makes your coding more efficient. Package that idea into an abstract and submit it to the regional conference's call for papers. If the technique is clever and useful enough, SAS programmers far and wide will end up using it in their own applications. How cool would that be?
- **Posters** As previously mentioned, poster presenters usually create both a technical paper published in the conference proceedings and a poster they present at the conference. The call for papers also requests ideas for posters. The Posters section is a good place for people who are not entirely comfortable with giving a presentation in front of a large group of people to begin their conference presenting career. They can build their confidence and presentation acumen with the generally smaller group of poster attendees and then move on to other types of presentations. Regionals usually schedule one or two times during the conference when all poster presenters are available at their posters to talk one on one with fellow conference goers who stop by to browse the posters.

- **Hands-on Workshops** Hands-on workshops are presented by top SAS programmers who can lead a group of people through a 90-minute class-like lecture. The instructor lectures, and then the attendees perform programming exercises on workstations. The exercises serve to reinforce and enhance the lectures. The hands-on workshop section chair invites the presenters in that section. So, if you have a particular expertise and can plan out a 90-minute workshop that includes exercises, you should contact the section chair well before the conference, in accordance with their Call for Papers submission guidelines. This is an excellent section in which to showcase your expertise in a particular facet of SAS while teaching the next generation of up-and-coming SAS programmers.

Giving presentations provides many personal benefits. It is a great way to hone your organizing and speaking skills, highlights your SAS programming knowledge, gets your work published, benefits the regional SAS community, and speaks well of your organization. So, it is essential for you to give presentations at regional SAS users groups in order to claim your place as one of the top SAS programmers.

The third main way of participating in regional SAS users groups is volunteering. Like all other SAS users groups, regionals are run by people from the SAS users community. The jobs of finding the meeting space, negotiating with hotels, organizing the conference sections, planning the schedule, creating the call for papers, pre-registering attendees, assembling the giveaways, giving conference goers their registration packet onsite, running the sections, overseeing the opening and closing sections, operating the mixers, and more are all done by SAS professionals like yourself who volunteer their time. It takes a small army of volunteers to plan and run a regional users group conference.

These are some of the typical volunteer positions available at regionals:

- **Executive Committee** This group is usually composed of previous conference chairs and conference chairs-to-be. The executive committee provides overall guidance for the regionals. They deliver continuity and history to upcoming conferences because they know what works and what does not work for running a successful conference. Executive committee members might take on tasks such as searching for hotels for future conferences, liaising with other regionals and with SAS Global Forum, and helping to publicize upcoming conferences.
- **Conference Chair** This person is responsible for all aspects of the conference[6]. The conference chair's main duties might differ slightly from regional to regional. However, they usually determine which sections will be held, pick the section chairs, select the giveaway items, oversee the food and room negotiations with the hotels, supervise the call for papers, manage advertising for the conference, attend to the budget, conduct the opening and closing sessions, and keep the conference on track during the days when it is being held. People who have participated as section chairs for several years and have expressed an interest in leading the regional are normally selected as conference chairs by members of the executive committee.
- **Section Chairs** These individuals run the various sections that make up the conference, such as Analytics and Statistics, Beyond the Basics, Business Intelligence, Coders' Corner, Data Presentation and Reporting, Hands on Workshops, and Tutorials. They are responsible for

writing a description for the section, vetting papers submitted by potential presenters, accepting papers and setting a schedule for when authors must submit both the draft and the final paper, scheduling the papers for various time slots at the conference, sending completed papers to the conference chair for publication in the proceedings, recruiting session coordinators, recruiting people to judge contributed papers, and making sure that the section runs smoothly throughout the conference. Section chairs usually have some expertise in the section that they run. For example, a statistician is more apt to run the Statistics section and a programmer likely to run the Coders' Corner section. At any given conference some section chairs will be first-timers who volunteered their services to the conference chair, while many others will be reprising their roles as section chairs of previous conferences.

- **Session Coordinators** Session coordinators help run the sections at a conference. They introduce speakers, pass out handouts, dim the lights, make sure the microphones are working, keep the presentations moving on time, and a myriad of other helpful tasks. Session coordinators are recruited before and during the conference, and it is easy to become one. You simply send an e-mail to the section chair or conference volunteer coordinator. Because most conferences can't get enough session coordinators, most volunteers are accepted. It makes sense to volunteer as a session coordinator in a section where you already expect to spend a lot of time. For example, if you intend to spend a lot of time in Data Visualization, that would be the ideal section to volunteer for. Repeat session coordinators who prove to be dependable often end up as section chairs in subsequent years.
- **Other Volunteering Positions** There are a variety of other volunteering positions available at regional conferences. Help is sometimes needed at the registration desk, at a book drive or other charity fundraiser, at an information desk, and even at a station that has maps of local eateries and sights. If you are flexible and want to help, the conference volunteer coordinator or conference chair is sure to plug you into a position where you can support the conference.

Make sure you volunteer to help run your regional's SAS users group conference. Start out as a session coordinator; work your way up to a section chair; then to a conference chair. These volunteer positions can provide positive career growth opportunities outside of your sphere of work and help you build your management and organizational skills. They will expose you to a wide variety of fellow SAS professionals who can give you great ideas on both SAS programming and management matters. These extracurricular activities look good on your resume. And, becoming an active part in a regional SAS users group raises your visibility as a top SAS programmer who is well connected with the regional SAS community.

Special Interest SAS Users Groups

Overview

Special interest SAS users groups are exactly what they sound like: users groups that focus on a particular industry or facet of using SAS. They provide settings where like-minded SAS professionals can gather to talk about best practices for using a specific type of SAS software or for using SAS in a particular industry. Consequently, the focus of the meetings is generally narrower and more concentrated than the focus of other users groups. But, that is exactly what the participants like because it meets their unique interests and needs.

As of this writing, here are some examples of special interest SAS users groups in the United States:

- Great Lakes JMP Users Group
- Insurance and Financial SAS Users Group (IFSUG)
- Pharmaceutical Industry SAS Users Group (PharmaSUG)
- SAS Retail Users Group (SRUG)
- SAS Utilities Users Group (SUUG)
- Virtual SAS Users Group (VirtualSUG)

Special interest groups vary in size, the number of meetings they hold, and the number of people who attend the meetings. Meetings can range from once a quarter to once a year; from in-person to virtual. The meeting venues vary from simple meeting rooms in members' organization to a full-size conference held in a major hotel. (That is the case with PharmaSUG.) As with other SAS users groups, these things are bound by the number of people who join, the number who volunteer, the amount of support from SAS, and the amount of money in the organization's coffers. Whatever the case, special interest group meetings are great places to learn more about your own particular SAS concentration from other practitioners. And, they offer venues you can use to showcase your own SAS programming knowledge, or your knowledge about the specific special interest topic.

How You Can Participate

Most special interest SAS users group meetings tend to be on par with local SAS users group meetings in terms of size and attendance. So, you can contribute by volunteering to present industry-related, or special interest software-related presentations in line with the group's interests. Connect with the special interest group via the links on the support.sas.com website. Determine whether a group meets your industry or SAS software interests. Then find out whether it is geographically accessible to you, or—if not—whether the group offers virtual meetings. Next, communicate with the users group contact and ask about the frequency of meetings, where the meetings are held, and how you can join. Express your interest in presenting a paper and propose a topic you think will be interesting to those who share the particular group's focus. (Of course, the topic should be something that you know well enough to present.) You will very likely get a positive response and find yourself presenting at a future meeting.

The Pharmaceutical Industry SAS Users Group (PharmaSUG) holds an annual two-and-a-half day conference very similar to the regional SAS users group conferences. There is a call for papers, a lineup of pharmaceutical industry-oriented conference sections, contributed and invited presentations, and hundreds of attendees at a major hotel or conference center. So, participating in PharmaSUG is very akin to participating in a regional SAS users group conference as discussed earlier. Access the call for papers on the PharmaSUG website and determine whether you can contribute a paper. Or contact a section chair and discuss the possibility of giving an invited presentation. Volunteer to help out at the conference as a session coordinator, paper judge, or in some other capacity. And, keep your eyes open for other possible ways to help out, such as working your way up to being a section chair. It is a big special interest group and there are dozens of ways to participate. However, the best ways are always the ones in which you either present or lead. So, be alert for those possibilities in particular.

SAS Global Forum

Overview

There is no way to overstate it: SAS Global Forum is the veritable Super Bowl, World Series, US Open, World Cup, NBA Championship, Stanley Cup, and so on, of SAS users group conferences. It is an annual conference similar in structure to the regional conferences, but everything about it is bigger. Much bigger! SAS Global Forum lasts more days; there are larger opening and closing sessions; there are more concurrent sections; there are more technical papers; there are more before and after conference training opportunities; there are more hands-on workshops; there are more SAS staff presentations; there are more opportunities to test drive SAS products; there are more SAS developers, managers, and subject matter experts on hand to talk to you. The meetups and kickback parties are larger; there is a huge SAS demo room; and there are thousands of SAS professionals eager to talk to and network with another. Dr. James Goodnight, CEO of SAS, and other SAS notables join the conference chair at the opening session to welcome attendees and set the theme for the three days of non-stop SAS information sharing. So, SAS Global Forum is an annual event that you should make every effort to attend.

So much of what was written about the regional SAS users group meetings, earlier in this chapter, applies to SAS Global Forum. This is the case because the first SAS Global Forum (previously named SUGI[7]) was held in 1976, and most local and regional conferences are modeled after it. SAS Global Forum typically has the following features:

- Opening session
- Concurrent technical paper presentations
- Hands-on workshops
- Posters
- Code clinic
- SAS support and demo area
- Kickback party and volunteers' party
- Meetups

The Opening Session and SAS Support and Demo Area are presented by SAS, while all other events are put on by the SAS Global Forum International users group. Refer to the section on regional SAS users group meetings for more information about the similar events at SAS Global Forum.

Here are some of the additional events usually held at SAS Global Forum:

- **Reception for Academic Attendees** This meeting provides a relaxed forum in which faculty and staff can share information about different career paths that involve the use of SAS with students and junior faculty. If you are an academic, whether you are standing in front of or sitting in a classroom, this is a forum you should definitely attend.
- **First-Timers' Session** The first-timers' session is designed to give first-time SAS Global Forum attendees an overview of the conference. Conference officials review the agenda and tell attendees how to make the most of their conference experience. This is an invaluable meeting for people unfamiliar with the events at this big conference.
- **Opening Night Dinner** The opening night dinner is held immediately before the opening session, and attendees usually go directly from the dinner into the opening session. The dinner is a great place to network with fellow SAS professionals, especially considering that many of them will be paper presenters, poster presenters, pre- and post-conference instructors, SAS employees, or just knowledgeable SAS programmers like you. Talk at the tables usually involves professional experiences using SAS, interesting programming techniques, and recommendations of must-see presentations.
- **Opening Session** Although all big SAS conferences have an opening session, it is worth talking about SAS Global Forum's opening session. Not only does the conference chair welcome attendees and get the conference off to a good start, but SAS notables are also on the program. Jim Goodnight, CEO of SAS usually gives a presentation. He often interacts with leading technical staff who man workstations and demonstrate new or improved features of SAS software. Other SAS notables talk about technical support, discuss the SASware ballot, give user feedback awards, recognize companies for their innovative uses of SAS, and interview the top business and technical staff or organizations who have leveraged SAS to make significant differences in the success of their businesses. The opening session is a great place to learn about significant capabilities of SAS and the strategic direction SAS is currently undertaking. But it is not strictly serious. The session usually ends with entertainment showcasing the musical, dancing, or singing talent of the city in which it is being held.
- **Technology Connection and Keynote Presentation** This event is held early in the morning of the first full day of the conference. The first part of the meeting gives conference goers a look at the innovative usage of some of SAS' more sophisticated software. Knowledgeable SAS product managers demonstrate capabilities of their software live onstage. They are available to discuss the software in-depth in the SAS demo room throughout the rest of the conference. After the technical demonstrations, there is a keynote presentation. Keynote speakers vary from year to year. It could be a technology guru, a sports star, an inspirational speaker, or a well-known humorist. Whatever the case, the keynote presentation always offers attendees something interesting to think about as they head into the main activities of the conference.
- **Lunch and Featured Presentation** The first two full days of the conference have optional catered lunches that you can pay an additional fee to attend. This is often convenient, because it saves you from having to go out and find a local place to eat. Instead, you can sit at tables with other SAS professionals and chat them up on all manner of programming topics. Toward the end of the lunches a featured speaker gives a presentation on a SAS topic. Topics can range from the serious to the humorous[8], and are always interesting. You can sign up for the

lunches when you originally register for the conference, or at the registration desk when you arrive at the conference, if space is still available.

The SAS Support and Demo Area of SAS Global Forum is especially noteworthy. The "demo room" is an exciting, football field-sized area housing hundreds of SAS booths and exhibits such as SAS code doctors, SAS community connections, the customer usage survey, ad hoc and scheduled software demos, the SASware ballot, Posters, a large SAS Publications area, a SAS Alliance booth, a SAS Education area, Super demos, and conference sponsor exhibits. There are usually several hundred people in the demo area at any one time, asking questions, talking to SAS technical staff, attending demos, meeting SAS Press authors, ordering publications, using social media, having the code doctors solve SAS programming problems, taking new SAS products for a test drive, learning about local and regional SAS users groups, and networking with one another.

The demo area is an ideal place to return to again and again between interesting technical paper presentations to learn more about SAS software, programming techniques, and publications. SAS typically holds the Customer Loyalty Reception early in the evening of the first day in the demo room. At the reception, complimentary food and drinks are served, SAS software developers and technical managers are on hand to meet with attendees and talk SAS, and sometimes SAS Press staff raffles off complimentary copies of attending authors' books[9]. The reception is another highlight of the conference and—as with so many of the conference's events—a terrific place in which to meet other SAS users.

How You Can Participate

SAS Global Forum is like the New York City of SAS conferences in that *if you can make it there, you can make it anywhere*—conference-wise. It really is the big time and attracts top SAS programming professionals from across the country and from around the world. So, if you are really serious about being a top SAS programmer, you absolutely must participate in SAS Global Forum.

Taking part in SAS Global Forum is similar to taking part in regional SAS conferences. You can give a presentation[10]:

- **Invited Presentations** Presenters are invited by the section chairs. The presentations are 50 minutes long.
- **Contributed Presentations** Presenters submit an abstract to the call for papers. They are 20 minutes long.
- **Coders' Corner** Presenters submit an abstract to the call for papers. They are 10 minutes long.
- **Posters** Presenters submit an abstract to the call for papers. Poster presenters usually stand before their posters to answer questions for two one-hour sessions during the conference.
- **Hands-on Workshops** Presenters are invited by the section chair. The workshops are 90 minutes long.

Because you are a top SAS programmer, you should be giving technical presentations at SAS Global Forum. So be on the lookout for the conference's call for papers. The SAS Global Forum call for papers is sent out by e-mail and posted on the website in the fall of the year before the conference. When the call for papers opens, submit your ideas for a contributed paper, coders' corner presentation, or a poster. If you have an idea for an invited paper or a hands-on workshop, contact the section chair via the e-mail address posted on the conference website and pitch your idea. Be sure to provide a well laid-out proposal so that the section chair can easily understand what you intend to present and its technical merits. Giving a presentation at SAS Global Forum is an ideal way to exhibit your SAS programming acumen to a wider audience of SAS professionals.

You can also participate by being one of the following:

- **Section Chair**[11] Appointed by the conference chair to organize and administer a section.
- **Session Coordinator** Enlisted by section chairs to help run a section during the conference.
- **Paper Judge** Enlisted by section chairs to judge presentations in a particular section.
- **Code Doctors** Recruited by the section chair to help attendees who drop by the code clinic with their real-world SAS programming problems.
- **Other volunteer** Opportunities such as staffing the book drive desk or charity fundraising desk are usually available.

These types of positions are ideal for SAS programmers who might not have an idea for a technical paper, who might not yet be comfortable delivering a paper, or who might have had their paper idea rejected. Don't think twice about it; volunteer! SAS Global Forum is run by volunteers like you who are committed SAS programming professionals. You can have a hand in helping to run a successful conference, raise your visibility in the SAS world, and make some enduring professional friendships.

If you find yourself not presenting a paper and not in a position to volunteer, you can still get a lot out of attending SAS Global Forum. Be sure to attend every presentation you can get to that will teach you something new about SAS or reinforce what you already know. Attend hands-on workshops, making sure to sign up early. Go to all of the social events and...be social! Haunt the demo room and talk to the SAS developers and project managers about new software and about your own organization's uses of the software it currently licenses. Take new SAS products for a test drive in the demo room. Pour over the new SAS publications and see which ones should be on your own SAS bookshelf. Attend the opening and closing sessions and the technology forum presentations. In short, get the most out of your three days at the conference so that you return to your organization brimming with ideas that will help propel you forward to being a top SAS programmer.

---

[1] Note that these screenshots illustrate the Users Groups web pages on support.sas.com that exist as of this writing. They are subject to change and might look different in the future. Nonetheless, you should be able to find a local SAS users group near you quite easily.

[2] An *abstract* is a one-to-three paragraph summary of the subject matter of the presentation, designed to give you an idea of the technical content that will be covered.

[3] The conference proceedings for regional conferences, PharmaSUG, SAS Global Forum, SUGI, and SEUGI can all be found online on the sasCommunity.org website.

[4] Some regional conferences also include 50-minute contributed papers, but 20-minute presentations are the norm.

[5] The author started his own SAS presentations career just this way. He submitted an abstract for a contributed paper and was shocked when it was accepted. It meant that he had to get up in front of a room full of people and give a presentation, which was something he did not normally do. He wrote the paper, gave the presentation, and won the Best Contributed Paper award for his section that year.

[6] Some regional SAS users groups have conference co-chairs instead of a single chair. Some have an academic chair and an operational chair.

[7] SAS Users Group International (SUGI) conferences were held annually from 1976 through 2006. There were 31 SUGI conferences, named SUGI 1, SUGI 2, and so on, through SUGI 31. In 2007, the conference was renamed from SUGI to SAS Global Forum. Conferences are now noted by the year they were held. The latest one as of this writing is SAS Global Forum 2013.

[8] For example, the author gave a very well-received presentation titled "It's Not Easy Being A SAS Programmer" as the featured lunchtime speaker at SAS Global Forum 2010.

[9] SAS Press authors are happy to sign copies of their books for the lucky winners.

[10] See the description of these types of papers in the section on regional conferences earlier in this chapter.

[11] This conference continues to evolve! As of this writing there are plans to replace Section Chairs with Content Advisory Teams (CAT) comprised of CAT Leads and CAT Members. There are also plans to replace Session Coordinators with Content Delivery Team Leads and Room Hosts.

Chapter 8: SAS Training and Certification

Introduction

How much do you actually know about SAS programming? Do you really have a deep understanding of SAS fundamentals and advanced SAS programming techniques, or do you have just enough SAS knowledge to keep you going? Some people know SAS from having taken courses in college or from learning it on the job. They pick up the essentials and learn some new programming concepts and techniques from interactions with their coworkers and through reading. They know just enough SAS to perform common programming tasks. There is nothing wrong with that; being self-taught is fine. In fact, Chapter 2 provides a list of essential SAS programming constructs you can use as a blueprint for self-teaching. Using Chapter 2 as a guide, you can wade through SAS online documentation and printed SAS publications at your own pace to plug the gaps in your knowledge. But, some people learn better in a structured classroom environment. So, what can you do if you would like to be taught SAS in a more organized and deliberate manner?

Fortunately, you don't have to invent your own curriculum for teaching yourself the intricacies of SAS programming. SAS Education offers a wide variety of training opportunities. They have classes ranging from the very basic to the very advanced, and in an assortment of training formats varying from classroom instruction to classes conducted over the Internet. Accordingly, you can take advantage of SAS' existing training infrastructure to educate both yourself and other SAS programmers in your company. SAS also features a half-dozen ways SAS specialists can earn a professional certification credential. Certification credentials denote that the individuals who earned them have acquired a quantitative amount of SAS knowledge recognized by SAS. It is a statement to both employers and clients that individuals with SAS certification are serious enough about their SAS knowledge to train to a specific level of expertise that enables them to pass the rigorous SAS certification exams.

This chapter discusses SAS educational opportunities that you can consider for yourself and for the other SAS professionals in your organization. The first section examines the many facets of training you can obtain from SAS Education. The second section provides an overview of SAS certification. While reading both sections, think about the ways you can organize SAS training and certification for yourself and for your coworkers. You need to take the initiative and assume the leadership role of bringing SAS education into your own organization. After all, you *are* the top SAS programmer.

SAS Training

The first part of this section provides an overview of SAS Education classes. It is designed to acquaint you with the types of learning opportunities that are available through SAS. After reading the overview, click **Training** on support.sas.com to get detailed and up-to-date information about specific class offerings. Determine which classes would help make you a better SAS programmer and put them on your "must attend" list. Then, work with your management to establish a plan for taking the classes over an agreed-upon time period.

The second part of this section discusses ideas for how you can organize SAS training sessions in your own organization. The information in the section, plus what you already know about your colleagues who use SAS, should give you enough material to be able to bring SAS training into your company.

SAS Education Classes

SAS Education offers a wide variety of SAS classes you can attend to increase your knowledge of SAS programming. Your gateway to these classes is the Training page on the support.sas.com website. When you access the Training page, you can browse the list of courses, investigate e-learning opportunities, determine where SAS training centers are located, view the training formats, and explore class discount opportunities. There are links for getting started with introductory courses, for beginning a new career using SAS, and for classes on advanced topics. So, the Training page should be your starting point for exploring the SAS classes SAS Education has to offer. Here is a screen shot of the web page as of this writing:

Figure 8.1: SAS Training Page

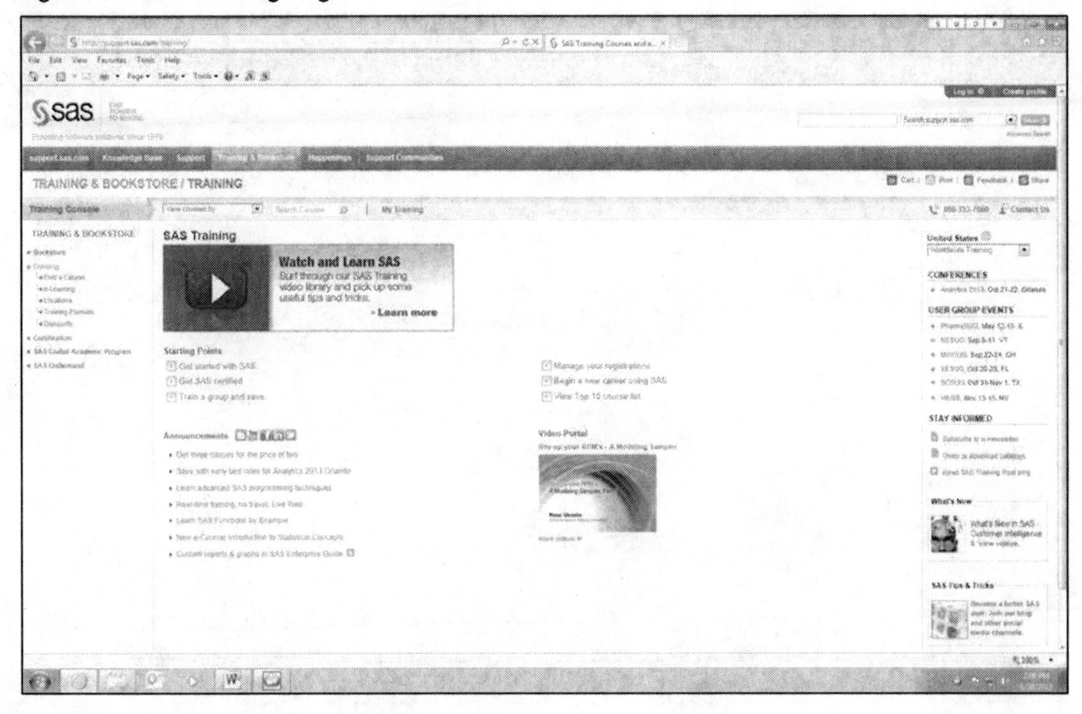

The Training web page offers many different ways for you to research all of the classes SAS offers. The most direct way is to click **Find a Course** and go from there.

Find a Course

The Find a Course link produces a web page that enables you to drill down on courses based on learning paths. Types of SAS learning paths include Foundation, Advanced Analytics, Business Intelligence, Information Management, Administration, and a host of SAS Solutions. Here is the Find a Course web page:

Figure 8.2: Find a Course Page

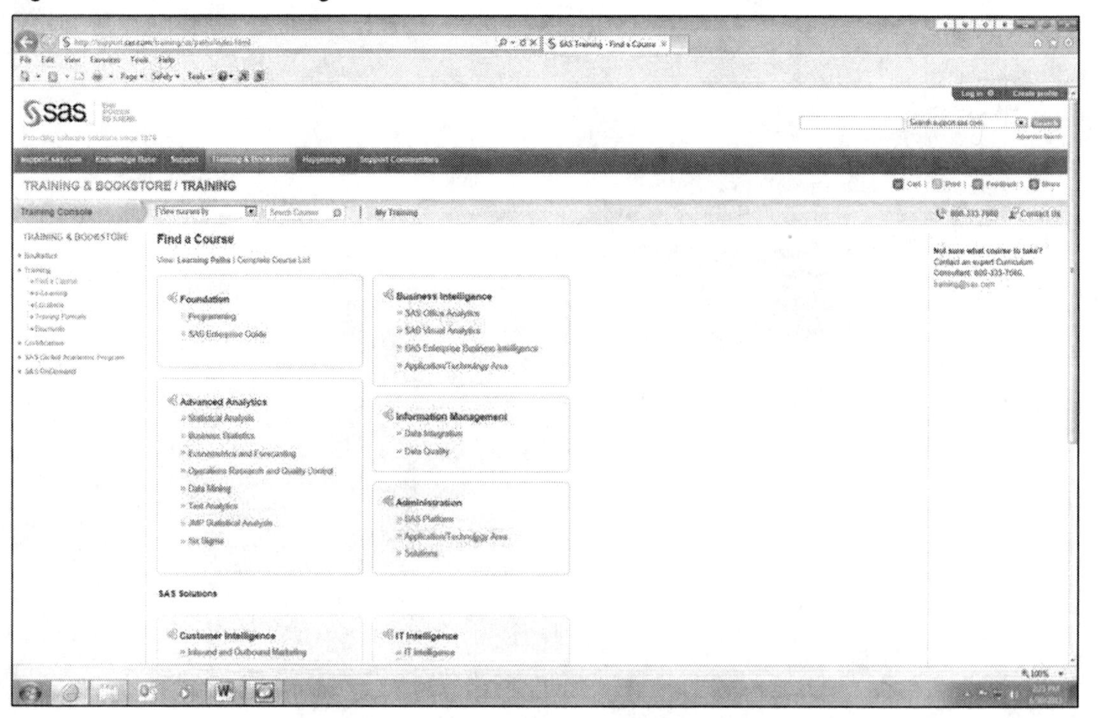

The learning path you choose will depend on the role you play in your organization and the SAS software it has licensed. More than likely, you will start with the Foundation learning path to learn the foundations of SAS programming. Click **Programing** to see the Foundation Programming web page:

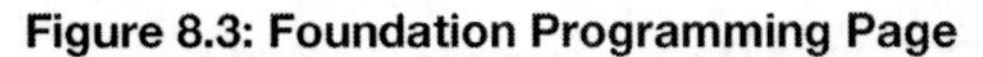

Figure 8.3: Foundation Programming Page

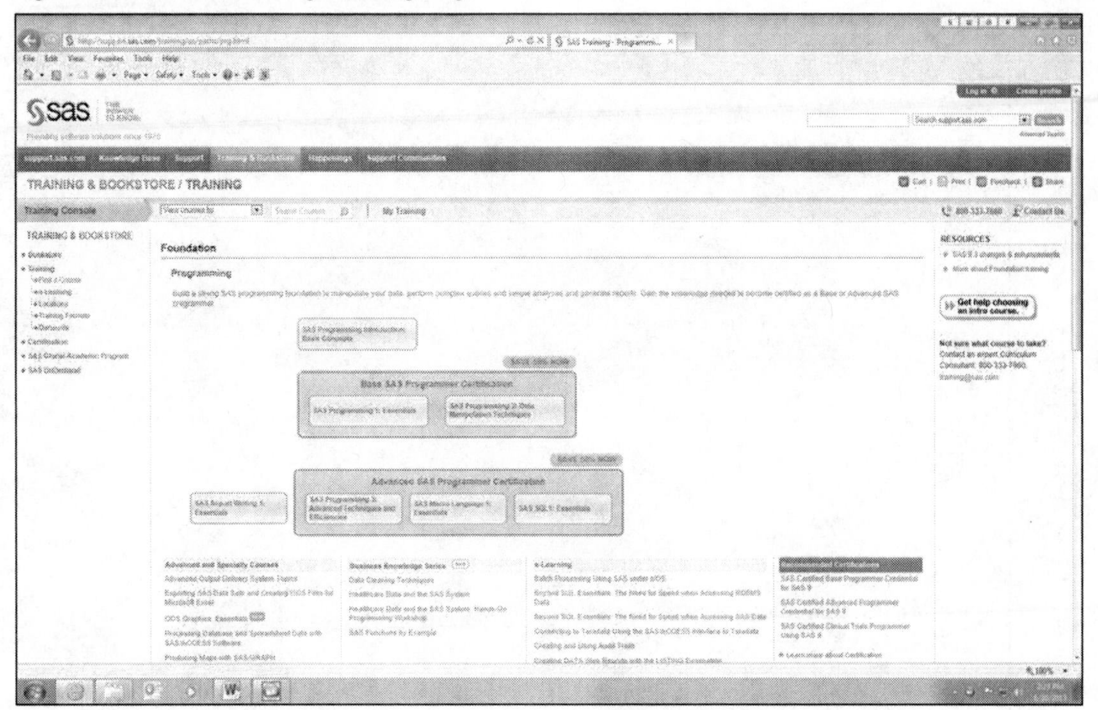

You can peruse the list of classes and find which are just right for you. The classes are arranged into logical groups so that you can take a series of classes to become proficient in a particular facet of SAS programming. Clicking on a particular class link takes you to web pages that provide an overview of the class, describe the prerequisites, and present the course outline. There is also logistical information, such as the formats the course is offered in (classroom, live web classroom, or e-learning), the duration of the course, the price of the course, the dates on which it is offered, and locations of the classes. Here is a screen shot for the *SAS Macro Language 1: Essentials* class:

Figure 8.4: SAS Macro Language 1: Essentials Class Page

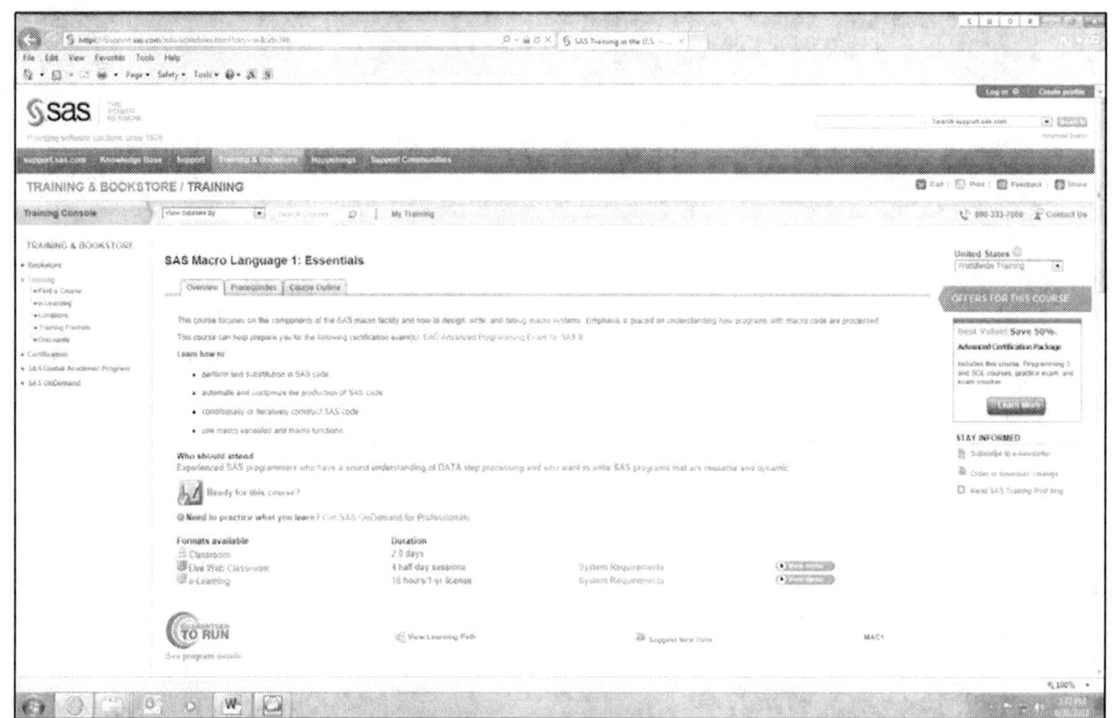

As you can see, the Find a Course web pages provide a very utilitarian way to identify the SAS classes that will further your SAS education.

Training Formats

SAS Education provides SAS training in a variety of formats, so it is easy to find one or more suitable for you and your coworkers. Here are the training formats currently offered:

- **Classroom** These SAS classes are taught in state-of-the-art training facilities located throughout the country. They normally offer hands-on programming exercises as well as lectures. Additional benefits of attending these classes are having the time away from the office to concentrate on the course, the individualized attention you get from an instructor when you have questions, and the ability to network with other SAS professionals. SAS has about three dozen SAS Training centers in the United States located in or near major cities.
- **Live Web Classroom** Live web classroom courses are taught over the Internet so that you can attend them from your computer at work. They are taught by SAS instructors who can answer your questions in real time. The main advantage to this format is that you do not have to travel and be out of the office, so you save travel expenses and are still accessible to your associates.

Another important benefit of the live web classes is that they are conducted in three-to-four-hour segments so that they don't completely take you away from your work duties. The sessions are also recorded and are available to class attendees for 20 days after the class ends. This is very beneficial if you have to miss all or part of a session because of other responsibilities or if you want to review part of the class to understand it better.

- **E-learning** This non-live format enables you to learn at your own pace by accessing course material over the Internet whenever you want to, 24/7. There are two types of e-learning courses:
    - e-courses offer training with quizzes, demos, and practices
    - e-lectures provide short, prerecorded lectures on specific topics

  Advantages to e-learning classes include not having to travel, being able to pick them up and put them down when convenient, and the ability to access the training from any computer that has access to the Internet.

- **On-Site Training** You can have SAS instructors teach classes in a training facility in your organization. This is ideal for training a group of programmers about a particular SAS topic. On-site training obviates the need for a group of staff members to be out of the office and eliminates travel costs. You can work with SAS Education to customize the class for your organization's specific business needs.
- **Mentoring Services** You can have a SAS instructor become an on-the-job coach to help you learn how to write SAS programs dealing with your own data in your own environment. The mentor can help you gain insights into your data and become more comfortable with the SAS programming language you use to process it. This can be a big advantage for SAS programmers who are still learning their craft.

This list of training formats is varied enough so that one or more of them is likely to meet your needs and your budget. You are the best judge of which learning environment is best suited for yourself.

Now that you know where to look for information about formal SAS classes, make good use of it. Peruse the list of classes and drill down on the courses that are pertinent to your job. Read the class outlines carefully and determine how much you stand to learn in a particular class. Make a list of the classes that would benefit you the most, paying attention to prerequisites and the best sequence in which to take the classes. Then, move forward with getting your training approved by your organization and scheduling yourself into some SAS classes. It's time to supplement what you already know with some formal instruction from SAS!

Organize SAS Training Sessions

Would you like to take on the responsibility of helping your colleagues and yourself become more educated and adept at using SAS? Of course you would! As the top SAS programmer in your organization, you realize the benefits that a well-educated population of SAS professionals can bring to your company. Programmers who truly understand the fundamentals of the SAS programming language can leverage it to complete assignments much better than those with only a cursory knowledge. A more educated SAS workforce can be more competitive because programmers have a wide variety of SAS tools they can bring to bear when working on real-world programming tasks. Well-trained SAS programmers can get assignments done quicker and with efficient SAS programs that use only the SAS constructs absolutely necessary to process data. So, SAS training is in your best interests and in the best interests of your organization.

The first thing you need to do is to assess what the SAS training needs are in your organization. What is the average level of SAS proficiency—beginning, intermediate, or advanced? How many programmers are at each level? What are the possible weaknesses in how staff is currently using SAS? For example, are your coworkers mainly using PROC PRINT and listing output instead of using more sophisticated SAS reporting routines such as PROC REPORT and the Output Delivery System? Could your associates use an introductory class to SAS and statistics? What about advanced topics such as efficiency techniques? Are your teammates making the best use of SAS Enterprise Guide? Take all of these questions—and more—into consideration to start getting a full picture of the training regimen that would be the most beneficial to the greatest number of SAS users.

The next step is to access the Training pages of support.sas.com and determine which classes are available. SAS Education offers an extensive assortment of class offerings, so you are sure to find any number of classes that intersect with your organization's needs. As previously discussed in this chapter, classes are offered in a variety of formats. There are classes held at regional SAS training centers. They require that you and your colleagues leave the office for a day or two to attend live, instructor-led training. There are live web classroom classes. These SAS Training instructor-led classes can be attended from the privacy of your office via the Internet. SAS provides self-paced e-learning classes. E-learning classes are taken over the Internet at your own pace. You can work through the material at your own speed for a specific number of months, stopping and restarting whenever it is most convenient for your busy schedule.

Another form of SAS training is on-site, instructor-led training. This is arranged by first purchasing SAS Training Points[1] on the SAS Education web pages. Then, you work with SAS Training to arrange for a particular class to be held in a training room in your organization. Such classes cost a specific number of SAS Training Points, depending on the type of class, its duration, and the number of students attending. The benefit of this type of training is that your colleagues do not incur travel expenses and still receive the advantages of working with a live SAS instructor. Of course, this requires you to have training facilities available in your organization. Many instructor-led classes have hands-on exercises, so they need workstations with SAS loaded onto them. You can iron out the details of which training room facilities are available when you approach your management with the idea of providing in-house SAS training.

Once you have a training plan and your management's approval, contact SAS Education and discuss your needs and ideas with the staff. They can help you arrange the classes that best meet your stated needs. SAS Education can provide pricing and invoicing information, and work out the fine details of arranging for an instructor if you are doing on-site or other custom training. They can even work with you to customize SAS training for your staff if none of the current training offerings quite fits your organization's needs. The SAS Training staff is invaluable in your effort to put together a successful in-house training plan.

When the training has been scheduled, work with the relevant IT managers to have the appropriate staff members booked into the class. Training is a serious matter, because it provides SAS programmers with the tools they need to become more effective in their work, which can make them more promotable. So, try to be as fair as possible in the selection process for the classes. Send those who are selected to attend the class an e-mail with pertinent information about the class. The e-mail should go out several weeks ahead of the class so that they can plan for it in their busy schedules. Be sure to copy their managers on the e-mail, so that they are on board with the dates and times their staff will be in training. Send attendees a reminder about a week out, and then on the morning of the class. Here is an example of a first-notice e-mail for a custom, in-house class taught by a SAS Training instructor:

All,

You have been confirmed to attend the following SAS class on **December 1st – December 2nd** from 8:30am to 5:00pm in the Building 5 computer training room:

> **Class:** SAS Report Writing 1: Using Procedures and ODS
> **Instructor**: Instructor 1, SAS Institute
> **Course description**: https://support.sas.com/edu/schedules.html?id=284&ctry=US

There is a waiting list for this class and we are paying SAS for 20 attendees. Let me know if you are unable to attend so I can fill the empty seat with another staff member.

The class schedule for both days is:

- 8:30 – 9:00 – Check in and handout of materials
- 9:00 – 12:00 – Instruction
- 12:00 – 1:00 – Lunch break
- 1:00 – 5:00 – Instruction

Note that the class will start promptly at 9:00am each day.

Thank you!

Sincerely,

Michael A. Raithel

Officiate at the SAS training so that the attendees know you are the organizer. This means you should introduce the SAS instructor in on-site classes on the first day of the class. Talk to the instructor beforehand and get a brief bio. When the class is about to begin, introduce the instructor and read the bio so that the class knows a bit more about who is teaching the class. Let the class know you are arranging classes, and ask them to provide you with feedback. Then, back off and let the instructor teach the class. Be sure to check back periodically to make sure there are no technical glitches with the equipment in the training room. If you are working in a large organization with support staff responsible for the training room equipment, let them know you want to be notified if there are any issues. You arranged the training and it is your responsibility to make sure it goes off without a hitch.

At the end of the year, write an e-mail to your upper management describing the SAS training conducted during the year. Provide the names of the SAS classes, the number of attendees, and a summary of the feedback you received from them. Solicit input from higher level managers for other SAS subjects they would like to have their staff trained in. Then, start crafting your organization's SAS training plan for the new year.

SAS Certification

Since its inception in 1999, the SAS Global Certification program has awarded over 50,000 SAS credentials to SAS professionals in 77 countries. The program recognizes SAS professionals who demonstrate an in-depth understanding of SAS software by awarding them with globally recognized credentials. Some of the goals of the certification program include the following:

- establishing a universal standard for measuring SAS software knowledge
- developing an internationally recognized certification program
- increasing the level of SAS software knowledge in the marketplace

As an individual, you benefit because you will have set yourself apart by earning the only globally recognized credential recognized by SAS. The SAS website lists the following benefits to individuals who become certified:

- increases your career opportunities and marketability
- enhances your credibility as a technical professional
- assesses your knowledge of SAS software
- allows you to earn industry validation for your knowledge

To earn a SAS certification credential, you must study the required curriculum for the credential and then take a certification test at one of the official test venues. You receive your test score directly after taking the test, so that you can immediately determine how well you did. When you successfully pass the test:

- You receive an e-mail with the credential logo attached to it. You can copy and paste the credential logo into your e-mails near your signature line to let people you correspond with know that you are SAS certified.
- You receive a diploma.
- Your name is posted to the online SAS Global Certified Professional Directory Listing.

Consequently, one and all can see that you have attained a high level of SAS expertise that is recognized by the SAS Institute.

There are currently seven SAS certification credentials within five SAS topic areas that you can earn:

- SAS Foundation
  - SAS Certified Base Programmer for SAS 9
  - SAS Certified Advanced Programmer for SAS 9
  - SAS Certified Clinical Trials Programmer Using SAS 9
- SAS Analytics
  - SAS Certified Predictive Modeler Using SAS Enterprise Miner 9
- SAS Administration
  - SAS Certified Platform Administrator for SAS 9
- SAS Data Management
  - SAS Certified Data Integration Developer for SAS 9
- SAS Enterprise Business Intelligence
  - SAS Certified BI Content Developer for SAS 9

As the top SAS programmer in your organization, the first two certifications under SAS Foundation seem like appropriate credentials to earn. However, do not discount any of the other certifications. The more you know, the more effective you can be at turning your enterprise's data into usable information.

The best place for you to begin your SAS certification journey is by clicking **Certification** under the Training & Bookstore tab on support.sas.com. That link produces the SAS Global Certification program home page:

Figure 8.5: SAS Global Certification Program Page

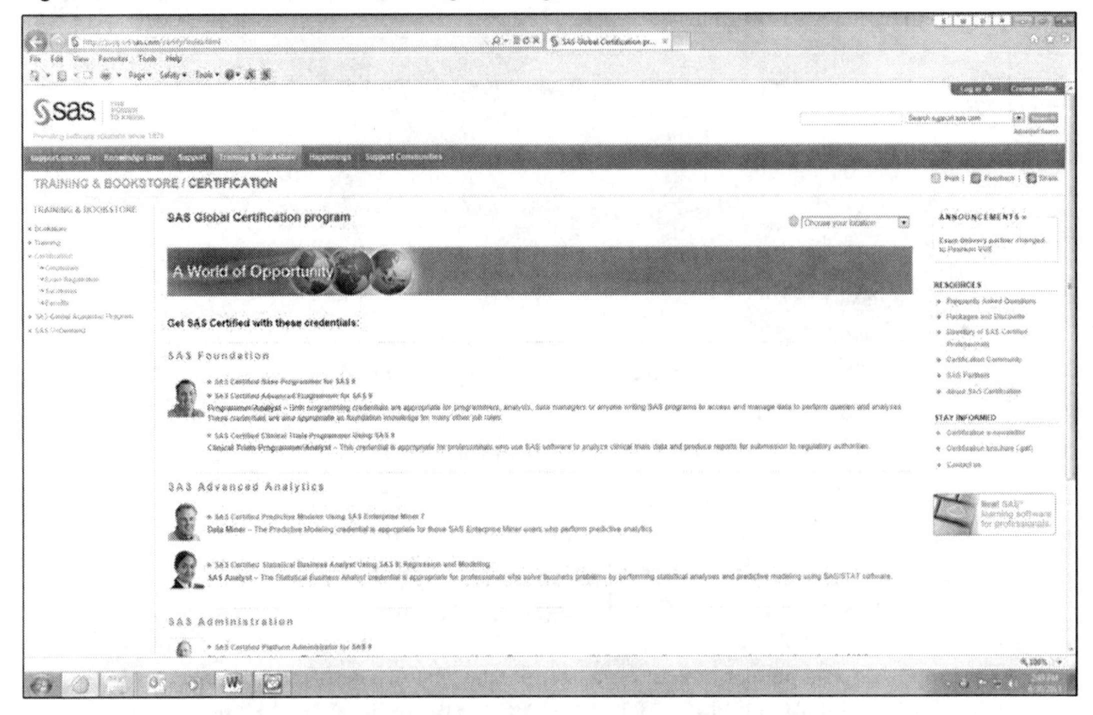

This web page is the starting point for in-depth information about each SAS credential, including an overview, an idea of the exam content, how to prepare for the exam, how to register for the exam, what to expect on exam day, and how you will be notified of the results of the exam. Clicking **SAS Certified Base Programmer for SAS 9** produces a page with the type of information you typically find for each certification credential:

Figure 8.6 SAS Certified Base Programmer for SAS 9 Page

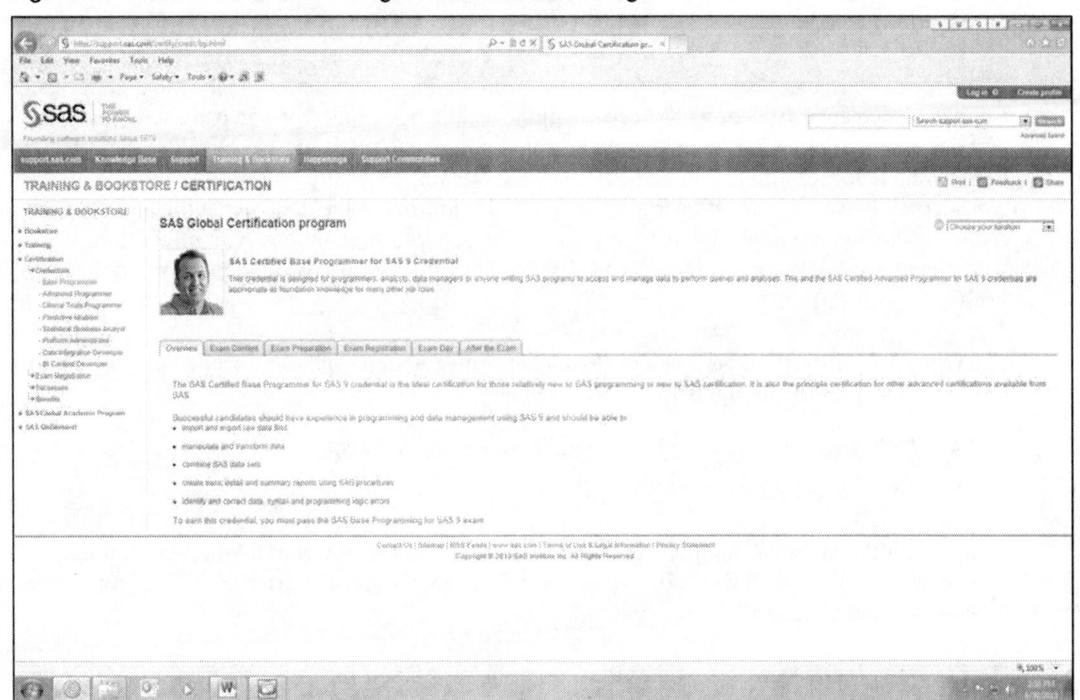

When examining these web pages, pay close attention to both the *exam content* and the *exam preparation* sections of the certification exam. All exams require an in-depth knowledge of SAS; sometimes more in-depth than most programmers are likely to have picked up on the job. The exam preparation courses can be invaluable both as a learning experience for things you are not familiar with, and for a review of SAS constructs and ideas you might not have used in a while. Review the exam content outline and determine the areas you are weak in. Then, see if some of the SAS courses recommended in the exam preparation section are suitable for plugging the gaps in your knowledge. For example, the SAS Certified Base Programmer for SAS 9 credential suggests you take the following two classes to prepare:

- SAS Programming 1: Essentials
- SAS Programming 2: Data Manipulation Techniques

Review the outlines of suggested courses in the SAS Education pages of support.sas.com to determine whether you need to take them. If so, schedule the courses in the appropriate training format and begin strengthening your SAS knowledge for the upcoming exam.

Conversely, if you are an experienced SAS professional with knowledge of the exam content areas, then the website recommends taking the SAS Certification Review: Base Programming for SAS 9 review course. Even if you have the core knowledge, you should consider the review class for the particular certification to ensure that you really are up to speed on all of the material you will be tested on.

If you are serious about earning a SAS certification, then it makes sense to take the practice exam for that particular certification. The certification website has a link that you can use to purchase the practice exam. Sign up for the exam and take it at a time and place most convenient for you. Review the feedback you get on your test. If your score is not where you want it to be, then consider taking some of the preparation courses. Or, go back to the documentation and research the sections where your knowledge is the weakest. It is better to find out how you might fare beforehand and take actions to improve your score than to not do as well as you can on the test itself.

The two SAS certification prep guides (available from SAS Press) make excellent supplemental resources for studying for the SAS certification tests:

- *SAS Certification Prep Guide: Base Programming for SAS 9*
- *SAS Certification Prep Guide: Advanced Programming for SAS 9*

The prep guides are thick compendiums full of comprehensive, relevant SAS programming knowledge. They are well worth the investment for preparing for the exams, and are great resources to have on hand afterward.

Here are the general steps that you can take to prepare yourself for SAS certification:

1. Determine which certification credential you want to earn.
2. Examine the exam content and determine how strong you are in your knowledge of what you will be tested on. Note topics you have not yet mastered so that you can learn them or brush up on them.
3. Review the suggested exam preparation regimen. Take any prerequisite SAS classes that plug in the gaps in the SAS knowledge you need to successfully earn the credential.
4. Take the practice exam and assess your score. If you are not satisfied with your score, study up on the topic areas you are weak in.
5. Sign up to take the test and take the test.
6. Enjoy the feeling of having earned another professional credit: a SAS certification credential.

Depending on the business your organization is engaged in, having a company-wide certification process for SAS programmers could be beneficial. If your organization provides contract programming for clients, then having a lot of certified SAS programmers might help you be more competitive on bids. Prospective clients would know that your company's staff went well beyond the norm and underwent the rigorous training and studying necessary to become SAS certified. They could feel confident your organization is willing to go the extra step to promote a high level of SAS expertise among its programmers. Such expertise would obviously translate into more effective usage of SAS to process data on their behalf and consequently quicker completion of project tasks.

Meet with your organization's senior management and discuss the possibility of promoting SAS certification as a standard within your enterprise. Perhaps the goal could be to ask all SAS programmers to consider earning the SAS Certified Base Programmer for SAS 9 credential. Those who did could be rewarded for their efforts at review time. Another possibility could be to request all of the senior programmers to earn the SAS Certified Advanced Programmer for SAS 9 credential. Keep in mind that the goal is to strengthen the SAS knowledge and professionalism in your organization. You don't want to alienate programmers by making certification an absolute requirement. So, work with senior management to craft a system of positive rewards for those who rise to the occasion and increase both their own and the organization's professional standing.

Promote the fact that SAS professionals in your organization have earned certification credentials by posting such information on the corporate intranet pages. For example:

XYZ Corporation takes the professionalism and training of its SAS programmers seriously. A significant number of our programming staff are SAS certified:

- 20 have earned the SAS Certified Base Programmer for SAS 9 credential
- 15 have earned the SAS Certified Advanced Programmer for SAS 9 credential
- 2 have earned the SAS Certified Predictive Modeler for SAS Enterprise Miner 6
- 1 has earned the SAS Certified Platform Administrator for SAS 9
- 3 have earned the SAS Certified Data Integration Developer for SAS 9
- 1 has earned the SAS Certified BI Content Developer for SAS 9

Because your colleagues have earned SAS certification credentials, it makes sense to tell the world, and especially your clients, by posting it on the corporate web page.

---

[1] SAS Training Points enable an organization to purchase training in bulk at a significant discount. Refer to the SAS Training Points Discount Program web page on support.sas.com.

Chapter 9: SAS Virtual Communities

Introduction

If you live in the United States, chances are very good that you spend a significant amount of your time in the virtual world. The *virtual world* is the world of the Internet with its web pages, e-mails, listservs, chat rooms, and social media outlets, as realized on workstations, laptops, netbooks, notebooks, tablets, and smart phones. Fifty-five percent of Americans use the Internet every day[1]. In a given month, the average American Internet user logs on to the Internet 57 times, spends about 60 hours logged on, and views over 2,600 web pages. About seventy percent of American Internet users take part in social networking. Considering that you are a computer professional, it is likely that you spend more time in the virtual world than the average person.

Not surprisingly, there are multiple SAS virtual communities that you can join to interact with SAS professionals throughout the country and from all over the world. These communities offer everything from web pages that are similar to Wikipedia to lively interactive discussion forums. You can simply sit back and consume the vast array of SAS educational material posted on the websites, the discussion forums, and the listservs. Or, you can become an active participant; creating articles and tips on wikis, and posting questions and answers to the discussion forums. It's up to you. But, since you are on your way to becoming a top SAS programmer, you know you should take your rightful place as a regular contributor and actively participant in SAS virtual communities.

This chapter provides an overview of the most prevalent and active SAS virtual communities[2]. It gives you enough information about each of them for you to get a sense of what they have to offer. Consider each of the virtual communities that are presented and begin planning how you can participate. Don't be hesitant. You have a lot to offer the worldwide SAS community, and your contributions will be appreciated. So, read up, determine which ones you want to participate in, access those communities, get a sense of what is going on, and then share your SAS expertise.

Discussion Forums on support.sas.com

It makes perfect sense that some of the most informative SAS programming discussions take place on the SAS support website. The SAS Support Communities web pages at https://communities.sas.com/community/support-communities provide access to over 20 web-based SAS discussion groups. Here is a screen shot of the SAS Support Communities home page as of this writing:

Figure 9.1: Example of SAS Support Communities Page

At this time, here are the available support communities:

- SAS Procedures
- SAS Macro Facility, Data Step and SAS Language Elements
- SAS/GRAPH and ODS Graphics
- ODS and Base Reporting
- SAS Web Report Studio
- SAS Enterprise Guide
- SAS Stored Processes
- Integration with Microsoft Office
- SAS Statistical Procedures
- SAS/IML and SAS/IML Studio
- Mathematical Optimization and Operations Research with SAS
- SAS Forecasting
- SAS Data Mining
- Text and Content Analytics
- SAS Enterprise Data Management & Integration
- JMP Software
- SAS IT Resource Management
- SAS Drug Development
- SAS in Health Care Related Fields
- Assistive Technology
- SAS Deployment

There appears to be a support community for practically every major kind of SAS interest. So, you shouldn't have trouble finding several communities that sync with both your own SAS interests and those of your organization.

When you navigate to the SAS Support Communities home page, you can view the recent discussions within each community by simply clicking on the link for the community (for example, SAS Procedures). Once you do, you can see a list of "Recent Discussions" with the various subject lines. Beneath each subject line, you can see the last time that dialog was updated (for example, 15 hours ago, 1 day ago, and so on). Simply click on one of the hyperlinked subject lines to navigate to a page that contains all of the postings for the subject. You can view the original question and all comments by subsequent responders. An especially nice feature of this forum is that some posters provide attachments so that you can read the actual SAS code or look at the output file.

Each one of the communities has links to relevant SAS documentation resources on the community's home page. For example, on the right side of the SAS Procedures community page, there is a boxed area titled "Resources for SAS Procedures." It has three links in it:

- SAS Procedures by Name
- SAS Language Reference
- Basic programming training course

So, if you come across a procedure (or facets of a procedure) that you do not understand, you can access one of the relevant links to find out more about it. How convenient!

Here is a screen shot of the SAS Procedures community web page:

Figure 9.2: The SAS Procedures Community Home Page

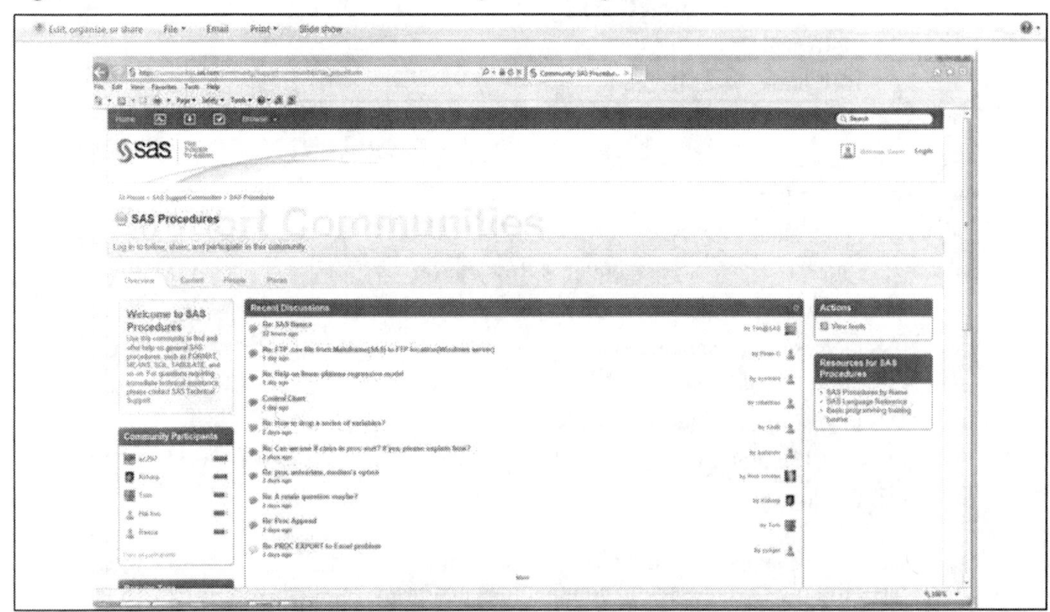

Here are some of the other items currently featured on the SAS Support Communities home page:

- Open a Track with Technical Support
- Message from the Community Manager
- Most Recent Content
- Most Recent SASware Ballot Ideas
- Community Participants

There is a lot to see on the Support Communities home page, so it is hard to characterize it all in words. The best thing for you to do is to open a web browser, navigate to the home page, and explore.

How You Can Participate in the Discussion Forums On support.sas.com

The easiest thing you can do to participate is to simply access the forums of interest to you and follow the discussions. Do this on a daily, biweekly, or weekly basis to stay current on what people are talking about. Read people's initial questions and think about how you would resolve their particular issue. Follow the answers posted by other SAS programmers and see if that is how you would have solved the problem. You can learn a lot from other SAS programmers' questions and the answers and solutions provided by other SAS professionals. This can also keep you on the leading edge of the SAS issues that are emerging as popular topics across many organizations. But, being a passive consumer of SAS information just isn't your style; you are going to want to contribute, too.

To participate in the discussions, you first need to create a SAS profile. This is easily done from the SAS Support Communities home page by completing an online form. Simply click **Login,** then in the New User section, click **Create**. After filling out the form and clicking **Create Profile**, you receive an e-mail with instructions to create your password and how to complete the profile activation process. When this is done, you are good to go and can post both questions and answers after you log on to the SAS Support Communities home page.

Here are some of the other things you can do while logged on:

- **Bookmark** – Link to and discuss content from around the web.
- **Discussion** – Start a conversation or ask a question.
- **Article** – Collaborate on an article.
- **Uploaded File** – Upload a file to share.
- **Poll** – Gather community opinion on a topic.
- **Status Update** – Share what you're up to.
- **Direct Message** – Send a private message to specific people.
- **Idea** – Create an idea for others to see and vote on.

As you can see, there are enough tools on the site to enable you to communicate in any way that is most natural for you. However, most of the interesting action takes place in the discussion forums.

sasCommunity.org

If you enjoy using Wikipedia to research general topics, then you are going to love using sasCommunity.org to research SAS topics. sasCommunity.org is a wiki[3] devoted to discussions and presentations of SAS. The wiki's home page states its mission this way:

> sasCommunity.org is an ongoing global community effort—created by SAS users for SAS users. This site is a place to find answers, share technical knowledge, collaborate on ideas and connect with others in the worldwide SAS Community.

You can see this is true from the moment you access the wiki at the following URL: www.sascommunity.org. The first thing you notice is the SAS Tip of the Day, which was contributed by a SAS professional like yourself. There is information about how to submit your blog, how to create an article, and how to submit a tip. You can browse recent blogs and articles, navigate through all of the SAS tips, and read forums and recent posts. All of this can be done via a few clicks on the home page of this very collaborative website.

Here is a screen shot of the sasCommunity.org home page as of this writing:

Figure 9.3: sasCommunity.org Home Page

Here are some of the main features of sasCommunity.org:

- **Blogs** This link provides access to blog postings by dozens of experienced SAS professionals on a variety of topics. Their informal writing styles and intelligent insights into how to use SAS make for great reading. You are sure to learn something new reading the blogs.
- **Sasopedia** This is a user-contributed knowledge base of information, code, and tips on using SAS. As such, it is ideal for researching various SAS components. Sasopedia is currently divided into four knowledge base areas:
  - **Language Elements** – Topics such as formats, functions and CALL routines, informats, options, and statements
  - **Procedures** – Information about various SAS procedures (for example, PROC APPEND and PROC GREPLAY)
  - **Products** – Factoids on SAS products such as SAS/ACCESS, SAS/IML, and JMP software
  - **Topics** – A plethora of SAS topics such as Best Practices, Descriptive Statistics, and the Program Data Vector
- **Popular Links** This section provides shortcuts to the most often accessed information on sascommunity.org, including these:
  - **Proceedings** With a few mouse clicks, you can access all of the SAS Users Group International (SUGI)[4] conference proceedings, the SAS Global Forum conference proceedings, the SAS European Users Group International proceedings, as well as the regional SAS users group proceedings.
  - **Tip of the Day** Many SAS users enjoy simply browsing through the SAS tips, which are written by experienced SAS programmers and vetted by sasCommunity.org reviewers.
- **Contribute** This section highlights how you can contribute to the wiki. Remember that the wiki depends on SAS professionals such as you to contribute content and help vet others' contributions. Be sure to read this section, consider how you can best contribute, and then take action and contribute to this popular SAS wiki.
- **Connect** You can use this page to connect to already-registered sasCommunity.org users, find SAS events, add your own SAS events to the wiki, or find users groups in your area. You can also provide feedback about the web site from this web page.

As you can see from the previous discussion, sasCommunity.org offers a diverse range of material about SAS programming, usage, and communities. It is a great resource to use to research SAS topics as well as to browse to find things you don't already know about SAS. As such, it should be on your web browser's Favorites list.

How You Can Participate in sasCommunity.org

SasCommunity.org is a great place to learn SAS programming techniques, find out about SAS software usage issues, and learn about SAS users groups. It is also a good venue for showcasing your SAS programming knowledge and increasing your visibility to the SAS world. But, before you can do either of these, you need to create an account. That is easily done by clicking **Create Account** and following the instructions. Once this process has been completed, you can log on to the website and begin adding or modifying content.

One of the easiest things to do is create a Tip of the Day. There are very good directions on the website that can guide you through the process. Once you complete your tip and submit it, a select group of sasCommunity.org volunteers vet the tip for accuracy. When they are done, they put it into the rotation of daily tips, and other users will eventually see your contribution. It is a lot of fun to access sasCommunity.org on a daily basis and occasionally see one of your own tips pop up as the Tip of the Day!

Another straightforward way to contribute is to create an article. Because you have expertise in a number of SAS topics, it should be relatively easy for you to write an explanatory article on one of them. Write the article offline using a word processor. Review the article for ease of reading, proper grammar, spelling, and accurate technical content. Then, log on to sasCommunity.org and contribute the article. It is quite a thrill to access the website and see your article in print, knowing it is available to SAS professionals from across the country and from around the world.

There are many other ways to participate in sasCommunity.org. If you do not want to add original content, then you can review existing content and provide updates to outdated postings or those rare articles that are not quite 100% accurate. The sasCommunity.org advisory board is always looking for energetic SAS professionals who are interested in "gardening" the content of the website. Gardening consists of helping keep the site attractive, useful, and technically accurate by making changes to existing content. There is a sasCommunity.org web page that describes gardening tasks. Access the gardening page and determine how you can use your own energy to improve the SAS wiki.

The sasCommunity.org advisory board, which is made up of many top SAS programmers, is always available to help you find the best way to participate in the wiki. If you have questions about the website, contact them. If you would like to participate but don't quite know how, contact them. If you have an idea for a new article or feature for the wiki, contact them. They will welcome your participation. And, you will be professionally richer for contributing to the premier wiki for SAS programmers created by SAS programmers.

SAS-L Listserv

SAS-L is a listserv[5] devoted to discussions of SAS programming and SAS software. SAS users post questions about real-life SAS programming problems, *how-can-I-do-this* type questions, and questions about SAS versions and features to the listserv. Other SAS professionals from all over the country, and from around the world, post answers to the questions. Some respondents post partial answers or simple tips on what the original poster could consider doing to resolve the issue. Others post complete solutions with code examples and references to relevant SAS documentation or SAS technical papers. Respondents often build on each other's answers and resolutions, deriving more and more elegant solutions to the programming problem as the discussion continues to evolve. Sometimes larger discussions ensue among the respondents so that readers gain extra knowledge of SAS and its behavior, which is well beyond the issues in the original programming question. SAS-L is a good vehicle for gaining a better understanding of the practical applications of SAS, as well as a forum for you to offer your own advice to practitioners.

SAS-L has been an active listserv since 1986. It is a user-run Internet mail list of professionals from the greater SAS programming community. The volume of e-mails received from the listserv can be high, with as many as 100 per day during particularly interesting discussions. There were 11,978 individual postings to the listserv in a recent year. That is a lot of SAS information being shared!

SAS-L conversations run the gamut from hard-core SAS topics to more lighthearted discussions. Some of the more popular threads in the past few years have had the following subject lines:

- Chance to Make SAS-L History
- Notes from SAS Global Forum
- My Vote for the Oddest SAS Feature
- Alternative for "GOTO" in a DO Loop
- Why We Need To Be Aware of Perl and R
- List of Retired SAS Procedures
- STRIP Function and Leading/Trailing Tabs
- Calculating Weighted Means
- SAS 9.3 Notes
- Why Query Doesn't Use Index
- Friday Humor

SAS-L is populated by many well-known SAS programmers. You can find postings from SAS Press authors, SAS Global Forum presenters, regional SAS users groups presenters, and even some SAS employees. Many people have become well-known SAS programmers through their participation in SAS-L. They consistently post relevant, insightful answers to technical questions, and contribute their ideas and perceptive observations to the discussions. Over a period of time, their constantly well-crafted responses and examples to original posters nets them a group of fellow subscribers who *always* read their posts. Consequently, such participants are on the rise to being top SAS programmers in terms of fame in the world of SAS programming.

That is exactly what *you* should be doing. You should be an active participant in SAS-L; reading questions and posting well-stated, concise, technically sound answers on the SAS subject areas you know about.

How You Can Participate in SAS-L

The first thing you need to do in order to participate in the SAS-L listserv is to subscribe to the list. This can be accomplished by navigating to the SAS-L password registration page: http://www.listserv.uga.edu/cgi-bin/wa?GETPW1 and following the instructions. The registration process asks you to enter your e-mail account and password. As previously mentioned, there can be dozens and dozens of e-mail messages sent from the listserv every day. So, you might want to use a separate e-mail account for your SAS-L messages.

Once you are fully subscribed to the SAS-L listserv, you begin getting the e-mails fellow subscribers send to the list. If you are new to listservs or to SAS programming, you might want to wait a while before making your first posting. Read the messages you receive and get a sense of SAS-L's own brand of *netiquette*. Take time to understand the ebb and flow of information between the participants. Learn new programming tips and techniques from the participants and try them out on your own SAS facilities to gain better insights into them. Save particularly interesting SAS-L postings to a special folder in your e-mail system so that you can refer to them in the future. In short; make the most of the constant flow of cutting-edge SAS programming information appearing in your Inbox every day.

SAS-L participants who read messages but do not post to the listserv are affectionately known as "lurkers." There is nothing wrong with being a lurker, especially if you are actively learning new things about SAS. But, you are not using the full potential of SAS-L by remaining a lurker. There are two ways to participate beyond being a lurker: posting a question and answering questions.

If you are stuck on a particular SAS programming problem, you can ask the list for tips and hints on how you might resolve it. Your e-mail to the listserv should have the following:

- **A meaningful subject line that gives an idea of the question being asked** Subject lines such as "Help!" or "Need SAS Advice" are not meaningful. Subject lines such as "Can I Create a SAS Data Set from PROC CONTENTS?" and "How Can I Change Style Elements in ODS PROC TEMPLATE?" are meaningful.
- **A clear statement of the problem** Don't confuse your readers with a lot of extraneous verbiage. Get right to the point and ask the question up front, early in your message. Provide pertinent background after asking the question. But, use sparse prose and only provide information needed to set up an understanding of your programming situation and what you need to know. Do not include superfluous material; it will serve only to dampen your overall message.
- **Sample SAS code** Include a sample of the SAS code you used that did not work the way you thought it would. This enables readers to better understand the issue you are describing.

- **SAS Log** Include the portion of the SAS log that illustrates the error, warning, or note you are questioning.
- **Version of SAS** Post the version of SAS you are using (for example, SAS 9.4 TS1M0). Newer versions of SAS have additional features and also fixes for reported issues. It is helpful for your readers to know which version you are having the issue with.
- **Specifics about your computing platform** Post the version of the operating system you are working on. For example, state whether you are working on Windows or Linux. Also, specify the version of your operating system such as Windows 7 or HP-UX Itanium 11.31.

Understand that SAS-L participants are professionals like you with job, family, and social responsibilities. They voluntarily provide information to posters when they can and at a pace that suits their commitment to other activities. So, be patient after posting your question to the list. You might get an immediate answer, or you might have to wait a while before somebody with the right expertise can address your particular question. Many people might respond, a few people might respond, and on rare occasions nobody might respond. More than likely, somebody will provide the answer you need or a hint on what direction you should take to resolve the issue. Do not use the listserv to do your job. Use it to ask for helpful information when you are stuck or need a hint on a programming direction to take.

When you do get an answer that helps, reply to the list. Let the respondent know his reply helped and how you were able to use the information provided. This lets the list know you have a solution and reinforces the proffered methodology for other people who had the same question that you had. And, it never hurts to say "thank you" to a person who goes out of his way to help you.

Posting responses to questions posed on SAS-L is a great way to share your expertise with the larger SAS programming community and to gain greater visibility. So, watch for questions you can answer or insights you can contribute to an ongoing thread. Craft your answers in simple, clear prose. Include example code and the pertinent parts of a SAS log if applicable. Make sure to explicitly address the poster's questions. A good way to do this is to repeat the poster's original question, followed by your reply.

Here is an example:

```
Dear SAS-L-ers,

Cindy posted this fine solution to SAS8832's query about sorting:

> This should get at what you're looking for:
>
> Proc Sort data=have out=check; by ID  date  descending Year; run;
> (assuming that Year is the variable that will tell you it's the
> latest).
> Proc print data=check (obs=20); run; (use a where statement if you
> have some that you know are going to be duplicates so you can see that
> it did, indeed, sort them in the order that you wanted).
>
> Proc sort data=check nodupkey out=sortedundup; by id date; run;
>
Cindy, nice solution!  I would only suggest adding the following to the second sort:

Proc sort data=check nodupkey out=sortedundup dupout=dupdates;
by id date;
run;

...simply as a double-check, so that the OP could browse through updates and see the observations that were
"discarded" as duplicates.  Just my QC paranoia talking here:-)

Cindy, best of luck in all your SAS endeavors!

----Michael A. Raithel
```

Here is another example:

Dear SAS-L-ers,

Quentin posted, in part:

> Again, by goal is to write ERROR: messages (and WARNING: messages) to
> the log, without having the word ERROR show up in my code. So that if
> someone searches a log for the word ERROR, they will only get a hit if
> there is actually an ERROR in the log. <s-n-i-p-!>
>

Quentin, very interesting problem and great write-up! I'm not sure how I would address the issue of continuations onto the next line; cleverer SAS-L-ers will undoubtedly address that. But, I do have a simple idea on how to address not having "ERROR" show up in the code. I would use this 1-2 punch:

1. Store the following SAS program as a file in one of your group directories, say: H:\group programs\errcodes.sas:

```
        options nosource;
        %LET E1 = Error:;
        %LET W1 = Warning:;
        %LET N1 = Note:;
options source;
```

2. %INCLUDE that program into all of your user programs and then use the Macros in your error messages, like this:

```
        %include "H:\group programs\errcodes.sas";

        data checkclass;
        set  sashelp.class;

        if age < 15 then put "&E1" "Too young for survey " name age ;

        run;
```

You will get your custom Error: message in the non-program portions of your SAS log.

Quentin, best of luck in all your SAS endeavors!

----Michael A. Raithel

While you might want to join in on all types of discussions on SAS-L, there are some topics you should definitely shy away from. Discussions about software other than SAS are best held on listservs that are oriented toward that particular software and not on SAS-L. Stay away from hot-button, non-software topics such as religion, politics, and social commentaries. Promoting products, goods, and services is also frowned upon on the SAS-L listserv. SAS-L-ers are pretty vocal when the occasional, uninformed listserv subscriber posts a message on a non-software topic and usually send chiding messages to the list. You are undoubtedly savvy enough to not make such a faux pas!

SAS E-Newsletters, RSS, and Blogs

The Happenings page (http://support.sas.com/community/index.html), which can be accessed via the support.sas.com home page, offers a link to the e-Newsletters page and the RSS and Blogs page. SAS staff communicate information about the latest features of SAS, changes to SAS, SAS programming and usage techniques, and a host of other topics in these free publications. They can keep you up to date on the latest developments from SAS, so you should subscribe to one or more of them. This section discusses the rich information sources the e-Newsletters, RSS feeds, and blogs provide.

E-Newsletters

The e-Newsletters page (http://support.sas.com/community/newsletters/index.html) presents a facility for subscribing to SAS newsletters on various topics. Newsletters are published periodically and sent directly to you via e-mails. You can also search the archives of previously published e-newsletters.

The following e-newsletters are currently available:

- **SAS Tech Report** – A monthly newsletter that covers news items related to technical support, training, publications, certification, and SAS events.
- **SAS Statistics and Operations Research News** – A quarterly newsletter aimed at informing statisticians, operations research specialists, econometricians, and data analysts of news and highlights related to SAS software they use.
  SAS Training Report – A monthly newsletter about training opportunities, conferences, and seminars in North America.
- **SAS Book Report** – A monthly publication discussing SAS books and online documentation.
- **SAS Global Certification News** – This quarterly newsletter covers news, events, and special offers for the SAS certified professional.

Subscription to all of the newsletters is free of charge. So, proactively subscribe to the newsletters that appeal to your interests and stay well informed on all things SAS.

RSS and Blogs

On the RSS & Blogs page, you can subscribe to RSS feeds, blogs, technical support listservs, and get information about how to subscribe to the SAS-L listserv.

The SAS RSS & Blogs home page is currently divided into three sections:

- **RSS Feeds** – This section is subdivided into feeds that provide content for the SAS customer and feeds that provide SAS corporate news and information. Here are some of the SAS customer feeds:
  - Content Highlights and Updates
  - Discussion Forums and Support Communities
  - SAS Statistics and Operations Research News

Here are some of the SAS corporate news and information feeds:

  - Press releases
  - Customer success stories
  - News features
- **Blogs** – This section enables you to subscribe to or view about a dozen blogs. Here are a few of them:
  - Key Happenings at support.sas.com
  - The SAS Dummy
  - The DO Loop
  - Peer Revue
- **Listservs** – Here, you can subscribe to the following listservs:
  - **SNOTES-L** – SAS Institute Technical Support uses SNOTES-L to distribute information about the availability of SAS Notes on the Customer Support Center website. This is a high-volume listserv, which sends mail on a daily basis with information about new and revised SAS Notes.
  - **TSNEWS-L** – TSNEWS-L is a means for SAS Technical Support to distribute technical information to customers. Information distributed on TSNEWS-L includes problem correction notifications, announcements about SAS hours and closings, and information about existing and new maintenance levels.
  - **SAS-L** – This is a user-supported listserv that you may choose to subscribe to.

Take advantage of the RSS & Blogs page by subscribing to the feeds of interest to you and that have a bearing on your work. Being notified of content changes and receiving newsletters is a good way of staying informed without having to do a lot of work searching around for changes. Let the feeds and listservs do the work of delivering the information to you as it becomes available.

---

[1] http://visualeconomics.creditloan.com/how-the-world-spends-its-time-online_2010-06-16/

[2] There are active SAS communities on social media websites such as Facebook, LinkedIn, and Twitter. However, they are changing so fast that it is not practical to characterize them in this chapter. Check them out and see what they might have to offer you.

[3] A wiki is a website that enables its users to create, change, or remove content.

[4] Members of the sasCoumminity.org advisory board did an outstanding job of completing the arduous task of scanning in hard-copy SUGI conference proceedings to make them available digitally. You can find proceedings from the first SUGI held in 1976 to present day SAS Global Forum conference proceedings.

[5] A listserv is an e-mail-based discussion forum that people can subscribe to. When one of the subscribers sends an e-mail to the list, the e-mail is automatically sent to all subscribers. Consequently, all subscribers can participate in listserv discussions either actively or passively.

Chapter 10: How You Can Become a Top SAS Programmer

Introduction

The earlier chapters of this book provide proven methods, ideas, strategies, and resources for you to use to get to the top of the SAS programming world. It doesn't matter whether you are a student studying programming, a SAS programmer with a few years of experience, or a SAS programmer with many years of experience; these ideas can work for you. But, you have to act. You have to take control of your career and work at becoming the best. It is time to consider how *you* can become a top SAS programmer.

This chapter highlights four main ingredients for working toward becoming a top SAS programmer. Read through the text and take the ideas to heart. Determine how you can shape and mold the ideas in this chapter to fit your own circumstances, and how they can work for you. Then, put them into action so that you become the top SAS programmer in your organization, a well-known top SAS programmer, a top-earning SAS programmer, or all three.

Make Use of All of the Resources Available to You

The most obvious resource is the very book that you are currently reading. This text is filled with hundreds of tips and techniques that are designed to help you successfully navigate your career toward your goal of being a top SAS programmer. Read the entire book and think very carefully about the advice. See how applicable the various techniques are to your own situation. Then, put those techniques into motion in your own organization or consulting practice.

Use the resources at work that are available to you. Take advantage of all internal training opportunities, whether they are related to SAS or simply related to the business environment you are working in. Participate in in-house seminars, webinars, and management meetings, especially those having a SAS focus and those having a data-related focus. Ask your management to sponsor you to become SAS certified, and then study the relevant material and take the certification exams.

Make use of resources outside of work. Ask your management to send you to SAS classes at a local SAS training center or at a local college or university. Try to attend local, regional, and international SAS users group meetings and conferences. Regularly read the SAS website, the SAS Support website, and other websites related to SAS. Read SAS books in your spare time. You might also consider the following:

- part-time consulting as a SAS programmer if it doesn't violate your organization's policies
- attending SAS product demonstrations, open houses, and webinars whenever possible
- writing articles for SAS users group newsletters
- learning SAS at home if your organization's SAS licensing agreement allows it
- writing a SAS Tip for support.sas.com
- reviewing a SAS book for SAS Press
- writing a SAS book for SAS Press
- participating in SAS discussions in virtual communities
- learning to use the SAS Business Intelligence software or the SAS analytics tools
- becoming the SAS Administrator for your organization

The key element is that you should be passionate about using the resources that are available to you both within and outside of the organization that employs you.

Ambition Is the Key Ingredient

Ambition is the key ingredient to becoming a top SAS programmer. You have to really want to achieve this goal and then work hard at becoming one of the best. This book provides the ideas for getting you there, but you have to supply the ambition and the work. You must continually set goals for yourself and then work toward achieving them. You cannot become complacent; you must always be tending your SAS programming career. So, get ready to unleash your pent-up ambition and become the top SAS programmer you were always meant to be.

A key to being successful is looking at what you do for a living as a "career" instead of as a "job." Consider the definitions of job and career:

- **Job** Something you do to make a living. A piece of work, especially a specific task done as part of the routine of one's occupation or for an agreed price.

- **Career** Something you do as your lifework. An occupation or profession, especially one requiring special training.

It is obvious that a career carries a deeper commitment to your life than a job. The point is that you must have a passion for the work you perform when programming in SAS. You must have a passion for using the SAS programming language to input, process, and analyze data. This comes about by identifying your work as a career. If programming is simply a job that you attend from 9:00 to 5:00, you will not do as well as if you regard it as lifework. So, get excited about your SAS programming prowess and ambitious about using it to rise to the top of your profession.

Publicity

It is a sad reality that many programmers who do their jobs, no matter how well, are not really recognized for their efforts. Sure, they get their requisite yearly raises and on-time promotions, but by and large, they do not get any "extra credit" for the many clever and innovative things they do. You will not go down that road, but heed some of the advice in these sage quotes:

The only thing worse than being talked about is not being talked about. – Oscar Wilde

Without promotion something terrible happens... Nothing! – P. T. Barnum

You must make sure your organization recognizes you for the work that you do. This means keeping your name out in front of your manager and your peers by making sure they know the innovative things you are doing with SAS software. You can do this by specifying what you are doing in staff meetings, in status reports, in special e-mails, and in hall conversations. Be creative and do things such as putting a footnote on all reports that reads something like this: "*Produced by The SAS Group – Call Us at Extension 3976*", so that every reader is aware it was your group that created the report. The rule of thumb is, "No good SAS deed must go unnoticed"!

Try to be clever and creative in keeping the things you are doing with SAS software in the limelight. Chapter 6, "What You Can Do in Your Own Organization," provides a lot of tried-and-true ways for getting and maintaining visibility. What you can really do will be influenced by the corporate culture of the organization you are working in. However, publicity for the many great things you do with SAS software is the key to becoming a top SAS programmer in your organization. So, you should constantly come up with and evaluate the best ways to let your peers and management know how you are contributing to the good of the organization using SAS.

Have an Aggressive Approach to Work

Your approach to work is one of the things that can help you to become a top SAS programmer. If you adopt a stimulus-response type of attitude, where you wait for an assignment, get it, complete it, wait for the next assignment, and so on, you are not going to go very far. Instead, aggressively seek out assignments, attack the business problems with your SAS programming skills, turn the assignments in early, and ask for the next assignment.

Be proactive. Do not wait for an assignment or other work to come to you; ask for more work or help your colleagues with their work. Work ahead of schedules. Don't wait until you finally get a request to extract data; do it now. Analyze newly delivered data early to see if there are any issues with it. Determine whether there are other deliverables your client needs or would use if they were available, such as new reports or data extracts, and create them *now*. Or, create a checklist that your clients can begin using to specify data sets, selection criteria, analysis variables, and so on, for future ad hoc requests. Be creative in being proactive!

Show leadership. Be willing to be a team leader or a manager. Such positions enable you to provide oversight to other SAS programmers. You can then help them to learn and use the latest SAS programming techniques and tools, so that your programming team is strengthened. Being a leader enables you to give assignments to other programmers. When you control the flow of assignments, you have a greater picture of what is happening in your organization and how the various pieces of a project fit together. Being in a leadership position raises your visibility and shows management you are a serious employee who is a valuable resource for the organization.

Innovate. When you are not in the midst of programming must-do-now assignments, take time to circle back and examine the existing processes to see how they can be streamlined. Can you organize programs and data for more efficient turnaround on ad hoc requests? Can you organize your existing programs so that you can better reuse them or more easily find pieces of reusable SAS code for future projects? Can you organize the data into a more cogent form so that it is more conducive to future analysis? Are there other SAS software modules you can license that would make it easier to access data? Be creative and determine exactly how you can make your programs, data, and processes more efficient and easier to work with.

Do not get bored with SAS! Make it a habit to learn something new about the vast world of SAS software every week. Revisit SAS programming language features that you already know and see if there are changes in recent releases that you can use in your programs to make them more efficient. Keep your SAS skills sharp by constantly looking for better ways to program. Learn SAS trivia. Make sure that you know more about SAS than any of your colleagues. Part of your being a top SAS programmer is knowing more about SAS than anybody else in your organization. Make it so!

Index

P

Q

R

S

X

Z

CPSIA information can be obtained at www.ICGtesting.com
Printed in the USA
LVOW05s1955031213

363718LV00013B/825/P

9 781612 901046